UNECE

Motorcycle Anti-Lock Braking System (ABS) for ASEAN Countries

World Forum for Harmonization of Vehicle Regulations (WP.29)
"How it works - How to join it" Series

UNITED NATIONS
Geneva, 2024

This publication is issued in English, French, Russian. United Nations publication issued by the United Nations Economic Commission for Europe.

ECE/TRANS/359

UNITED NATIONS PUBLICATION
Sales No.: E.24.II.E.30
ISBN: 978-92-1-003355-8
eISBN: 978-92-1-106938-9

Photo credits:
Cover: Adobe Stock Photos,
p. 8 Adobe Stock Photos, p 10 Bernard Frost.

Acknowledgements

The publication "**Motorcycle Anti-Lock Braking System (ABS) for ASEAN Countries**", aims to increase motorcyclist safety through the focus on crash avoidance technologies. To promote the efforts made by the UNECE stakeholders in the development of UN Regulation 78/ UN Global Technical Regulation (GTR) No. 3 for Motorcycle ABS. The publication was produced, initiated, and prepared by the UNECE Transport Division in cooperation with contracting parties and several stakeholders. The Transport Division wishes to express its sincere thanks to all those who contributed to this publication.

A special thanks goes to those who took part in the consultation process:

The main author:

Mr. Bernard Frost **United Kingdom**

Other Contributions

Mr. Azzaharin bin Allias **Malaysia**

Ts. Muhammad Firdhaus bin Ibrahim **Malaysia**

Automotive Engineering Division, Road Transport Department **Malaysia**

The Malaysian Institute of Road Safety Research (MIROS) **Malaysia**

Nguyen Anh Duc **Viet Nam**

Viet Nam Register - Motor Vehicle Technical Safety Inspection Department **Viet Nam**

Engr. John Steven M. Magboo **Philippines**

Standards Conformity Division, Bureau of Philippine Standards **Philippines**

The Towards Zero Foundation **United Kingdom**

Mr. Walter Nissler **UNECE**

Ms. Melissa Archer **UNECE**

United Nations Road Safety Fund **(UNRSF)** for the financing of activities

Contents

List of tables

List of figures

Executive Summary

In August 2020 the UN General Assembly announced a second UN Decade of Action (2021-2030) with the ambition to reduce global road traffic casualties by 50% by 2030. World Health Organization (WHO) data indicate that 92% of road traffic deaths occur in low to middle income countries. WHO also note that the Southeast Asia region accounts for 28% of road traffic deaths. In 2016 the Association of Southeast Asian Nations (ASEAN) developed a Regional Road Safety Strategy that, like the UN Road Safety Strategy, contains 5 pillars of action in a safe-system approach. The adoption of motorcycle anti-lock braking technology is identified within the pillar on safer vehicles.

Motorcycle anti-lock braking has been available for many years but, for a long while it was only available on a select group of large motorcycles. The efficiency of the technology has been the subject of extensive research for many years. In 2009, the European Commission engaged the Transport Research Laboratory (now TRL Ltd.) to undertake a study of measures that could be introduced to improve the safety of this vulnerable group of road users. Amongst the technologies considered was anti-lock braking. The report reviewed a number of individual published studies and concluded that that the technology had the potential to significantly reduce casualties. However, at that time the benefits associated with its application on low-capacity motorcycles was less clear. The cost of the technology was unclear and, at its highest estimated cost, benefit to cost analysis for smaller motorcycles tended toward a negative in this early period.

The European Commission revisited the question in 2020 asking TRL to conduct a further review. By this time manufacturers had reduced the size of the anti-lock braking components, reducing issues for packaging the technology on a motorcycle. The cost of the technology was also more easily assessed. The Transport Research Laboratory review reported from stakeholder engagement that suggested fitting costs of between €100 and €150.

Together the individual studies for the European Union indicate a safety benefit associated with the technology and, as a result more refined cost estimates, a positive benefit to cost ratio. The European Union has introduced legislation that mandates the technology on motorcycles of 125 cm^3 or greater.

Motorcycles are a significant method of transport in the Southeast Asian region. It is reported that in 2022 the number of motorcycles in-use in Viet Nam was 72 million with as many as 6 million new motorcycles being registered each year. It is understood that motorcycles with engines with a capacity of less than 250 cm^3 are dominant.

It has been reported that motorcyclists underutilise their braking systems, even in emergency conditions. This is particularly true of the front brake, and it is thought that a fear of locking the wheels is a significant contributor to this response. Research has indicated that even a brief period of wheel lock, less than 0.5 of a second, can cause instability and a fall. With the front brake providing 70 – 80% of the braking capability, underutilisation will result in significantly increased stopping distances/less than optimal speed reduction, and riders who systematically avoid its use are unlikely to be prepared to use it in an emergency situation.

Vavryn and Winkelbauer, (2004), showed that for motorcycles that were not equipped with anti-lock braking, on average, an experienced rider could achieve decelerations of 6.6 m/s^2 whereas a novice rider achieved average decelerations of 5.7 m/s^2. However, after 6-hours of training using motorcycles equipped with anti-lock braking, experienced rides achieved an additional 1 m/s^2 in deceleration while novices achieving an improvement of 2 m/s^2; almost matching the ability of experienced riders – 7.8 m/s^2 and 7.7 m/s^2 respectively. These differences represent significant braking improvements that could prevent collisions or, if avoidance was not possible, much reduced impact speeds.

If anti-lock braking systems are to be encouraged, or legally required, it is necessary to provide clarity on the performance that the technology is required to deliver. Braking regulations, made under the auspices of the United Nations Commission for Europe's World Forum for Harmonization of Vehicle Regulations, include detailed performance requirements for anti-lock braking systems. With UN Regulation No. 78 being widely used by manufacturers as the global braking standard for motorcycles (or, in some cases, the equivalent UN Global Technical Regulation No.3) National Administrations have a ready-made text to use as part of any initiative to advance the penetration of the technology into the vehicle fleet.

The ambition of the ASEAN Regional Road Safety Strategy could be delivered by regulatory measures. However, it is noted that the ASEAN region has also been engaged with the Organisation for Economic Co-operation and Development (OECD) to enhance good Regulation principles. These principles include the development of a Regulatory Impact Assessment (RIA) to establish the most effective method of delivery. An RIA will consider both regulatory and non-regulatory options.

Regulatory impact assessments are a commonly used tool and, in addition to informing best policy decisions, they also provide the necessary detail to use in a consultation documents – whether pursuing a regulatory or non-regulatory path.

With the ambition for anti-lock braking being a regional one, there is an opportunity for the ASEAN countries, perhaps under their Regional Road Safety Strategy, to develop a benefit to cost study that could be a precursor for an RIA, and that could be relevant across the region with only detailed changes necessary for the individual countries. The WHO has developed tools to overcome limited, or incomplete, data which provides an opportunity to profile the casualty situation in individual countries using a standardised model. This would provide a strong policy document that Administrations could deploy.

Whether or not a regulatory approach is pursued, or whether additional policy levers are employed, the case for using the UN Regulations to provide the technical requirements remains powerful. The Regulations provide options for UN type-approval, national type-approval, or a self-certification approach. They also lend themselves for use in a regulatory regime or to establish standards that could be employed in incentive schemes.

Introduction

According to the WHO Global Status Report on Road Safety 2023[1], 1.19 million lives were lost globally in 2021 due to road traffic injuries; this represents a fall of 5% compared with the estimated 1.25 million fatalities in 2010. These latest data represent 15 fatalities per 100,000 population. The number and the rate of road traffic fatalities and injuries vary across the world, with 92% of deaths occurring in low and middle-income countries and with lower income countries sustaining a fatality rate of 21 per 100,000 population.

Road traffic injuries are the twelfth leading cause of total deaths globally and are the leading cause of death for children and young adults in the 5–29 year-old age group. Globally, half of fatalities are among vulnerable road users (pedestrians, cyclists and motorcyclists).

The United Nations Economic and Social Commission for Asia and the Pacific (ESCAP) notes the progress made globally in reducing road casualties but also recognises the high road casualty rates in the Asia Pacific region with South-East Asia accounting for 330,222 fatalities; 28% of global fatalities. The rate of road fatalities in the South-East Asia region is the highest in the world at 28 per 100,000 population.

The UN General Assembly confirmed the ambition to provide access to safe, affordable, accessible and sustainable transport systems for all, including in the area of vehicle safety. In the second Decade of Action the target of preventing 50% of road traffic deaths as part of the plan of action is reaffirmed with a target date of 2030.

Globally, motorcyclists represent 21% of road traffic deaths but in South-East Asia, this figure rises to 48%. In some countries in the region, such as Cambodia and Thailand, rider fatalities are as high as 73% and 74% respectively. Since 2011, the global number of registered motorcycles has grown by 175%, with the number in the South-East Asia Region increasing by 273%.

Motorcyclists are a particularly vulnerable road user group and generally are overrepresented in road casualty data; this makes them a high priority for mitigation. Injury mitigation technology for motorcyclists is limited; primarily to crash helmet and protective clothing performance. Recent changes to UN Regulation No. 22 (22.06) have increased the minimum protection available from crash helmets. The capability of protective clothing has increased with the inclusion of body armour and, most recently, air bags integrated into protective jackets. However, protective clothing with advanced safety features remains expensive and its uptake may be low, even in high income economies, and it may also be less attractive in hot and humid regions.

The limitations associated with protective clothing demand that attention is given to crash avoidance measures. In many countries, to increase their conspicuity, motorcyclists have adopted a voluntary code to ride with their headlights switched on, and in some regions this is now compulsory. In addition, motorcyclist clothing is often produced using light coloured material that helps the rider to be seen by other road users.

However, one of the most effective crash avoidance technologies available to date is motorcycle anti-lock braking. This technology was first introduced on motorcycles in the late 1980's but its efficiency, size and cost have reduced significantly since those early systems came to the market. Based upon a study of the effectiveness of this technology, the European Union mandated anti-lock braking for motorcycles from 2016. That requirement extends to all new motorcycle models with an engine capacity greater than 125 cm^{3}.

To increase motorcyclist safety, there is an urgent need to democratise safety globally through the application of a motorcycle anti-lock braking systems as standard; empowering consumers to purchase a motorcycle with the technology. The objective of the Motorcycle Anti-Lock Braking System (ABS) Advocacy Campaign in the ASEAN Countries, is to support the adoption of motorcycle anti-lock braking and to strengthen national capacity to reduce motorcycle traffic fatalities and injuries in the ASEAN region.

---

[1] References Global status report on road safety 2023. Geneva: World Health Organization; 2023. Licence: CC BY-NC-SA 3.0 IGO.

Anti-lock Braking

Typically, a motorcycle is equipped with two individual brake controls, one for the front wheel and one for the rear. Different arrangements of braking controls can be found. Larger motorcycles tend to use a hand control on the right hand side of the handlebar for the front brake and another, right-foot operated control for the rear brake. The rider has to balance the application of these two individual brakes to achieve the best stopping performance.

Some motorcycles have combined braking systems, sometimes called advanced braking systems, where one of the controls actuates both a front and rear brake while the other actuates just the front brake. An alternative, sometimes found on low performance motorcycles, is to use two hand controls, as you might expect to see on a bicycle. An example of this arrangement (in Europe) is the Peugeot Elystar. Here, the right-hand brake lever operates the front brake – with anti-lock braking. The left-hand lever operates the brakes of both the front and rear wheels with anti-lock on the front wheel only.

It has been observed that the fear of locking the front wheel can cause motorcyclists to underutilise the front brake and, in some cases, not use it at all. With the relative braking performance of the two individual brakes being heavily weighted toward the front brake (70% - 80%) it is clear that stopping distances will be extended significantly or, if a collision is unavoidable, vehicle speeds will not reduce to the full potential, if the front brake is not fully exploited. The presence of an anti-lock braking system gives the rider the confidence to use the brakes to their fullest extent and therefore achieve the best possible stopping performance.

Frictional forces are present at the interface between a vehicle's tyre and the road surface. Those forces act in the longitudinal plane to provide acceleration or braking, and in the lateral plane to provide directional control and stability. If a wheel stops rotating (100% wheel slip) those forces reduce and the stabilising gyrostatic effect of the wheel is lost. When this happens, particularly in the case of a front wheel, it is likely to result in the rider falling. Research indicates that, for a motorcycle, and particularly for the front wheel, even a brief wheel lock (less than 0.5 second) can result in loss of control and a fall.

Wheel slip is the difference between the distance covered by the periphery of the tyre and that covered by the wheel's axle, or centre. When those distances are the same there is zero wheel slip and when the wheel is locked (no rotation) there is 100% wheel slip. Best stopping performance is achieved when there is a significant amount of wheel slip. Anti-lock braking systems aim to maintain that wheel slip at a high level but below that at which wheel lock can occur, assisting a rider to maintain control of the motorcycle and secure shorter stopping distances.

Figure 1 **Wheel Speed Sensor and Exciter Ring**

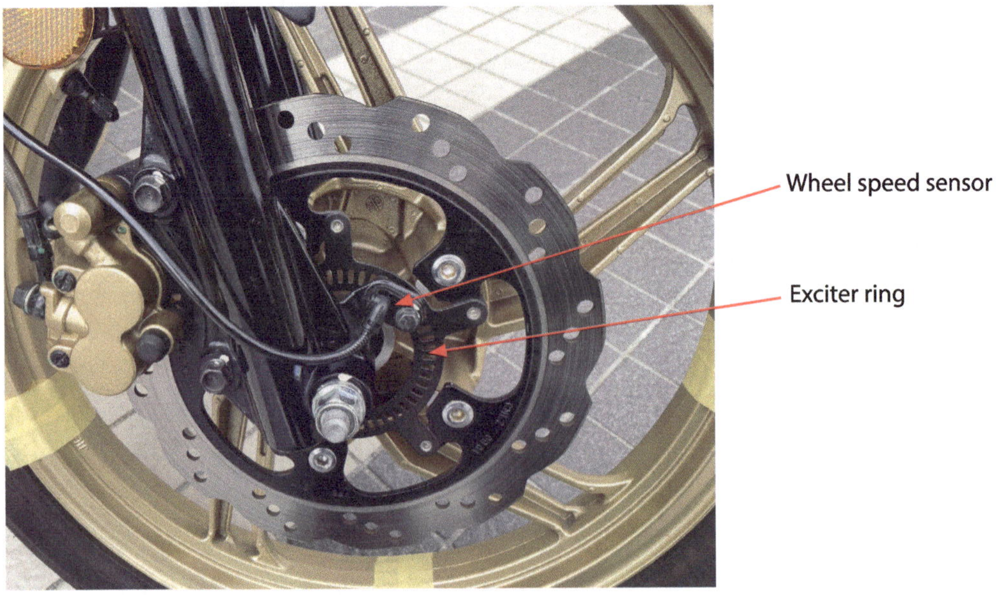

Source: Bernard Frost

The anti-lock braking system monitors the rotational speed of the road wheel and, sensing when the wheel slip is reaching a critical level, the brake is released and then re-applied. This action is repeated many times a second as the system seeks to keep the braking forces as close to the maximum as possible. Information about wheel speed is provided by a wheel speed sensor. The sensor comprises an electromagnetic coil that is located in close proximity to a toothed (exciter) ring mounted concentrically to the wheel centre. As the exciter ring passes by the sensor it causes a change in magnetic field which generates a change in voltage. The electronic control unit (ECU) deduces changes in wheel speed from the way in which this voltage changes.

Wheel lock can occur as a result of the rider applying excessive force on the brake control, even on a homogenous surface. It can also occur when a reasonable braking force is demanded but the wheel passes from one surface to another offering a lower level of adhesion, from dry to wet surfaces, lane markings etc. To be effective, an anti-lock braking system must be able to react to these surface transitions and optimise the braking force.

One - Channel and Two - Channel Anti-lock Braking Systems

A one-channel ABS system will have a single sensor, providing "one channel" of information to the ECU. Equipping only a single wheel with anti-lock protection delivers safety benefits but this should be used on the front wheel because, the risk of a fall is greater if this wheel locks under braking.

It follows that a two-channel system will have a sensor on each wheel to provide additional information to the ECU and permit individual modulation of the two brakes. This arrangement will provide the greatest confidence to the rider, allowing them to concentrate on controlling the motorcycle rather than balancing the braking.

Retro-fitting an anti-lock braking system.

Anti-lock braking systems are developed for inclusion as part of the manufacture of the motorcycle. At this time no manufacturer is known to offer a retrofit solution.

As described above, the function of the anti-lock system is to release and reapply pressure in the hydraulic line of the braking system many times a second, in response to the deceleration of the road wheel. The response and recovery of that wheel to a braking force will depend on the brake demand but may also be influenced by other factors, the inertia of the wheel unit for example. In addition to practical issues associated with the post-manufacture installation of an anti-lock system, the manufacturer of the anti-lock system would need to be involved in confirming the suitability of individual components for the intended motorcycle. The critical nature of the braking system also means that an Administration considering promoting this approach would need to have certainty about the integrity of a system before endorsing its use.

In the absence of agreed standards, retrofitting is not recommended.

Policy Considerations

The focus of this report is to support the adoption of motorcycle anti-lock braking into the legal framework of the countries that comprise the Association of Southeast Asian Nations (ASEAN). As with the UN Road Safety Plan, the ASEAN Regional Road Safety Strategy[2], published in 2016, sets out an agenda under 5 pillars:

1. Road safety management,
2. Safer roads and mobility,
3. Safer vehicles,
4. Safer road users,
5. Post-crash response.

Pillar 3, safer vehicles, lists five specific objectives, two of which refer to motorcycles. One is to develop approaches for safer carriage of passengers, and the other is to improve safety by promoting the adoption of anti-lock braking and, also, electronic stability control.

The inclusion of motorcycle anti-lock braking in the ASEAN Region Road Safety Strategy indicates that, from a policy perspective, the desire across the region to bring forward measures to increase motorcycle anti-lock braking is penetration already agreed.

Separately, the Organisation for Economic Co-operation and Development (OECD) works with ASEAN to develop best practice regulatory procedures. The ASEAN-OECD "Good Regulatory Practice Network", promotes regulatory reform and good practice among policy makers.

OECD support is provided at both a regional and individual country level. Country level support includes;

- regulatory strategy and oversight,
- good regulatory practices – regulatory impact assessments (RIA),
- stakeholder engagement and evaluations - regulatory delivery,
- international regulatory cooperation,
- regulating in the age of emerging technologies, and
- the application of behavioural insights.

OECD published work associated with the ASEAN member states include:

- Administrative Simplification in Viet Nam (2011)[3], which explores how Viet Nam can rapidly bring about the full potential their administrative simplification programme and introduce a range of regulatory instruments.
- Implementing Good Regulatory Practice in Malaysia (2015)[4], which examines Malaysia's early experience of implementing RIA to support evidence-based rule making.
- Regulatory Impact Assessment in the Philippines (2020)[5], which provides an assessment of the state of RIAs in the Philippines and a review of the new RIA requirements and oversight body.

The use of Regulatory Impact Assessments (RIA) to determine the delivery of policy objectives is a widely established practice. The OECD describes an RIA "a systemic approach to critically assessing the positive and negative effects of proposed and existing regulations and non-regulatory alternatives."

---

2 ASEAN Regional Road Safety Strategy Jakarta, ASEAN Secretariat, October 2016

3 OECD (2011), Administrative Simplification in Viet Nam: Supporting the Competitiveness of the Viet Namese Economy, Cutting Red tape, OECD Publishing. http://dx.doi.org/10.1787/9789264096646-en

4 OECD (2015), Implementing Good Regulatory Practice in Malaysia, OECD Publishing, Paris. http://dx.doi.org/10.1787/9789264230620-en

5 OECD/ADB (2020), "Regulatory Impact Assessment in the Philippines", OECD, Paris.

OECD analysis shows that "conducting an RIA within an appropriate systematic framework can underpin the capacity of governments to ensure that regulations are efficient and effective in a changing and complex world."

As indicated above, an RIA should consider regulatory and non-regulatory methods to address the policy concern. It is likely to include a "do nothing" option – leaving it to the market. The RIA process should establish the costs and benefits to different stakeholders and analyse the economic, environmental and social impacts of potential solutions. The following points are indicative of an RIA assessment:

- Defining the policy concern
- Identification of the mitigation
- Cost and Benefits of the mitigation
- Consideration of options for providing mitigation
- Impact on business (including small to medium size enterprises (SME))

An RIA should provide not only the evidence to support policy choices but also the evidence, and the detail, to support consultations with stakeholders regarding the policy delivery. The OECD provides information regarding the RIA process and procedure on its website[6].

The inclusion of motorcycle anti-lock braking in the ASEAN Road Safety Strategy, and the working relationship with OECD regarding regulatory best practice, opens an opportunity for regional cooperation regarding and RIA.

The policy concern, identification of mitigation, and the policy options for delivering that mitigation are likely to be very similar across the region. There will be national differences in the costs and the benefits and, perhaps, on the impact on small and medium enterprises. However, a regional RIA could establish a template for these factors that could be used at a national level. Cooperation on the production of a regional, but flexible, RIA could simplify the adoption of the technology among the ASEAN member states.

Policy Concern

According to the WHO Global Status Report on Road Safety 2023, 1.19 million lives were lost in 2021 due to road traffic injuries globally. Road traffic injuries represent the twelfth leading cause of total deaths globally with a fatality rate of 15 per 100,000 population, and the number one cause of death for children and young adults of the age group 5–29 years. The number and the rate of road traffic fatalities and injuries vary across the world, with 92% of deaths occurring in low and middle-income countries and with lower income countries sustaining a rate of 21 per 100,000 population.

Globally, motorcyclists represent 21% of all deaths and in the South-East Asia Region, this figure rises to 43% of all deaths. In some countries, such as Cambodia and Thailand, rider fatalities are as high as 73% and 74% respectively. The number of registered motorcycles has grown with the South-East Asia Region seeing an increase of 273%.

In the WHO South-East Asia (SEA) Region, 330,222 people were killed in 2021. The rate of road fatalities in the region is the highest in the world at 28 per 100,000 population.

Figures from the ASEAN Regional Road Safety Strategy show that, in 2014, Malaysia suffered 6,674 road traffic fatalities of which 59% were riders of motorcycles. It was reported that this toll came with an associated cost equivalent to 1.6% of GDP.

More recently, the ASEAN Motorcycle ABS report[7], prepared by the Malaysian Institute for Road Safety Research (MIROS), showed that between 2015 and 2019 the number of registered motorcycles in Malaysia rose by 16 percent and represent 46 percent of the vehicle fleet. The report records that approximately 4,000 Malaysian motorcyclists received fatal injuries in each year from 2015 to 2019. This represents approximately 64 percent of all recorded road traffic fatalities.

MIROS report an estimated cost of road traffic fatalities at 2 billion USD.

In the ASEAN Regional Road Safety Strategy, Viet Nam reported 8,671 road traffic fatalities in 2015, with an estimated cost equivalent to 2.45 percent of GDP. More recently the total number of fatalities has fallen; approximately 8,000 in 2016 reducing to 6,500 in 2022.

---

6 Regulatory Impact Assessment | OECD iLibrary (oecd-ilibrary.org)

7 Malaysian Motorcycle ABS Report MC+ABS+status+report.pdf (squarespace.com)

In 2015 their fleet comprised 46,792,091 vehicles of which 95 percent (approximately 44,000,000) were motorcycles. Figures for 2022 indicate a significant growth in the vehicle population to 99.9 million, of which approximately 72 percent (72 million) were motorcycles.

Recent data show that road traffic collisions have shown an encouraging fall in recent years, from over 20,000 in 2016 to approximately 11,000 in 2022. This may be reflective of the reduced road traffic fatalities. However, motorcycles remain a significant proportion of these collisions with approximately 72% of all collisions involving motorcycles in 2022.

Again, for the ASEAN Regional Road Safety Strategy, the Philippines reported that the vehicle fleet comprised 7,690,638 units with motorcycles accounting for 55.3 percent. Data for 2021 indicate the fleet has grown to 14 million registered vehicles with 62 percent being motorcycles. Of the motorcycle population, 80-85% are of a capacity of 150cc or less. It was estimated that in 2013 there were 10,379 road traffic fatalities, accounting for 2.6 per cent of GDP.

The MIROS study used WHO data which showed that, for 2016, there were 10,012 road traffic fatalities, of which motorcyclists represented only 4.7 percent.

During a "Consultative Meeting on Motorcycle ABS", (Kuala Lumpur; November 2023), representatives of the Philippines were only able to report data for the Metro Manilla region. This showed 613 fatal collisions with 295 (48 percent) of those being motorcyclists. To support robust policy decisions, it would be necessary to undertake further work to establish and analyse national road casualty data.

It is clear that analysis has been done across the ASEAN region to quantify the financial effect of road casualties on the economy. It is not clear if this quantification has used a common methodology or national models. Nevertheless, the presence of such data provide encouragement for the development of an RIA and this quantification should be explored to assist in either a regional or national impact assessment.

Mitigation: Effectiveness of Motorcycle Anti-Lock Braking Systems

In 2009, the Transport Research Laboratory (now TRL Ltd.) conducted a study for the European Commission aimed at identifying new safety measures for Category L vehicles (motorcycles)[8]. This study included an assessment of the benefits of anti-lock braking systems. As part of that research, T. L. Robinson, M. McCarthy, M. Pitcher, T. Gibson and C. Visvikis, conducted a review of relevant published research. That comprehensive review remains relevant today. The following observations are taken from that review.

Vavryn and Winkelbauer, (2004). This Austrian Road Safety Board paper aimed to qualify and quantify how anti-lock braking systems improve the braking performance of riders during an emergency braking manoeuvre. The subjects comprised both experienced and novice riders. They found that for motorcycles that were not equipped with anti-lock braking, on average, an experienced rider could achieve decelerations of 6.6 m/s^2 whereas a novice rider achieved average decelerations of 5.7 m/s^2. After 6-hours of training using motorcycles equipped with anti-lock braking, both rider groups improved their braking performance with novices achieving similar decelerations as the experienced riders – 7.8 m/s^2 for the experienced rider and 7.7 m/s^2 for the novices.

Vavryn and Winkelbauer recommended that the technology should be made mandatory on all motorcycles.

Sporner and Kramlich, (2000). Conducted in-depth studies of 610 motorcycle collisions. They found that in 65 percent of all collisions between a motorcycle and a car the motorcyclist had used the brakes to try and avoid the collision. They concluded that anti-lock braking would have delivered a positive influence on around 55% of the 610 collisions. In a 2001 paper, the European Transport Council quoted the Sporner and Kramlich research. They suggested that anti-lock braking could reduce the number of casualties by at least 10 percent.

Gwehenberger et al (2004). Reviewed 200 serious collisions to evaluate the effectiveness of anti-lock braking. Their work confirmed that the technology shortened stopping distances and prevented riders from falling. They also observed that the technology reduced the effort for the rider, particularly during emergency braking events. They predicted that the technology could avoid between 8 and 17 percent of all serious collisions. They also noted that to gain the full advantage of the technology, riders should receive training regarding its operation and use.

---

[8] Evaluating the impact of possible new measures concerning category L vehicles (Transport Research Laboratory) report_new_measures_l_category_en.pdf (europa.eu)

Kebschull and Zellner (2007 & 2008). Reviewed European data for 921 collisions (2000-2004), and 900 collisions from the research of Hurt et al (1981) for the USA. These data included both motorcycles and mopeds. This study found that anti-lock braking had a low effectiveness and that the technology costs were high compared to other "low effectiveness" options.

In 2020 TRL Ltd. conducted a further study for the European Commission[9]. This study looked again at the effectiveness of the technology and was aimed at evaluating the application of anti-lock braking on motorcycles within the European Union to smaller capacity; any motorcycles with an engine capacity greater than 125 cm^3.

Again, the following observations are taken from that study.

Teoh (2011). Compared the rate of fatal collisions in the USA between 13 motorcycle models equipped with anti-lock braking and the same models without the technology. Each of the models chosen would fall into the high-performance (>35kW) motorcycle category in Europe. He found that the rate of fatal collisions per 10,000 miles ridden was 37 percent lower for motorcycles equipped with anti-lock braking. The TRL report notes that Teoh published a further paper in 2013 in which the effectiveness was revised to a value of 31 percent.

Rizzi et al. (2016). Looked specifically at the risk of permanent medical impairment. Again, they used a comparison of motorcycles models with and without anti-lock braking. They found that the technology reduced hospital admissions by 47 percent and that the severity of injuries that did occur was reduced by 15 percent for impairing injuries and 37 percent for severely impairing injuries.

The combined effect of the technology in reducing the probability of a collision and reducing the severity of a collision if it did occur, reduced the risk of an impairing injury by 67 percent and a severely impairing injury by 55 percent.

Rizzi, Matteo Strandroth, Johan; Kullgren, Anders; Tingvall, Claes ; Fildes, Brian. This paper, "Effectiveness of antilock-brakes (ABS) on motorcycles in reducing crashes, a multi-national study" (23rd Session of the Conference on Enhanced Safety of Vehicles), explains their evaluation of the effectiveness of Antilock-Brakes (ABS) on motorcycles in reducing real-life crashes.

They noted that previous research mostly analysed large displacement motorcycles. They used police-reported crash data from Spain (2006-2009), Italy (2009) and Sweden (2003-2012) in an attempt to analyse a wide range of motorcycles, including scooters, and compare countries with different motorcycling habits.

The effectiveness of motorcycle anti-lock braking in reducing injury crashes ranged from 24% in Italy to 29% in Spain and 34% in Sweden. The minimum effectiveness with 95% confidence limits was 12%, 20% and 16%, respectively. The reduction of severe and fatal crashes was even greater, at 34% and 42% in Spain and Sweden, respectively. The minimum effectiveness was 23%-24%.

The overall reduction of crash involvement with ABS-equipped scooters (at least 250cc) in Italy and Spain was 27% and 22%, respectively.

Based on Spanish data only, anti-lock braking on scooters with at least a 250cc engine was found to reduce the involvement in severe and fatal crashes by 31%.

They concluded that, "at this stage, there are more than sufficient scientific-based proofs to support the implementation of anti-lock braking on all motorcycles, even light ones". However, further research should be aimed at understanding the injury mitigating effects of motorcycle ABS.

Policy Delivery Options

National/Regional Administrations have a number of options available to them when considering policy. In the case of changes to vehicle standards these could include:

- Do nothing. Rely on consumer demand, and manufacturer promotion, to increase the density of the technology in the market place.
- A consumer awareness initiative to raise awareness of the benefits of the technology.

---

9 MIS065-Examination-of-the-mandatory-fitting-of-an-anti-lock-brake-system-and-a-supplemental-combined-brake-system-to-L3e-A1-subcategory-motorcycles.pdf (trl.co.uk)

- Provide an incentive to accelerate consumer uptake, e.g.,
- A subsidised scrappage scheme for older vehicles.
  - A reduction in purchase taxation or annual registration tax.
  - Work with other parties, such as the insurance industry, to reduce the ownership costs of vehicles equipped with the technology.
- Regulate the fitting of the technology on new vehicles for a target group; novice riders, or those that are statistically the most vulnerable.
- Regulate for the fitting of the technology on all new vehicles.
- Regulate the fitting of the technology on all new vehicles and prohibit the use of vehicles that are not so equipped after a "reasonable" period of time.

Summary of mitigation options

1. Do Nothing.

   This approach relies on an existing consumer demand coupled with a desire, and capability of the manufacturer, to fulfil that demand. MIROS report a strong consumer demand for anti-lock braking to be fitted to all motorcycles (78.5 percent of 1200 respondents). However, the survey was conducted among 5 of the 10 ASEAN countries and 70 percent of the respondents were from only 2 of those.

   It would be necessary to determine more robust data for each of the ASEAN countries in order to assess the likely success of a market led solution across the region.

2. Consumer awareness.

   Consumer awareness programmes can be in the form of:

   - Information campaigns, radio, television, advertising, etc.
   - Detailed product assessment programmes - ASEAN NCAP is an example.

   Both are designed to encourage the consumer to make the right choice when buying a new product.

   Started in 2021, MIROS has an initiative for a motorcycle product assessment programme. While such programmes communicate directly with the public, much of their success comes from being able to differentiate between the safety potential of one manufacturers product over another; influencing each manufacturer to at least match the other. This requires a detailed specification for the assessment programme and a repeatable and reproducible methodology to differentiate performance.

   If successful, consumer information programmes can deliver improvements in the market more quickly than regulation. They can promote competition between manufacturers and stimulate consumer demand. They can also avoid the lead-in time associated with legislation.

3. Incentives

   Incentive schemes offer another way to bring technology forward. This approach imposes a cost onto the national Administration and it would be necessary to consider carefully the cost and the benefits to establish the nett value. In the case of motorcycle anti-lock braking, the maximum benefit may be gained by targeting the incentive toward the vulnerable group, for example, owners under the age of 25. Targeting an incentive can be difficult but could, for example, offer a purchasing incentive to a vehicle owner under the age of 25, who has a motorcycle that is at least 5-years old, and which has been registered in their name for at least 2-years before the scheme's start date. A condition of the scheme would be for the old motorcycle to be delivered into a scrappage scheme.

   A different type of incentive could be gain the cooperation of the insurance industry to discount premiums to young riders if their motorcycle has anti-lock braking on at least one wheel. Similarly, users of these motorcycles could be offered discounts on any annual registration fee.

4. Regulation

   A regulatory approach provides certainty, but it may also be relatively slow in delivering the full benefit. Generally, a regulatory approach will require new vehicles that are brought to the market to satisfy new technical standards. In general, the introduction of new regulations include a "lead time" for implementation that allows for industry to adjust to the new requirement, this lead time can be anything from 12-months to 60-months and may be determined as part of the regulatory consultation.

   Once implemented, the penetration of the technology into the national fleet is dependent upon the number of new vehicle sales, and the number of older technology vehicles leaving the fleet. For example, in 2022 Viet Nam had 7.2 m motorcycles in circulation. Figures for the first 9-months of 2023 indicate that 4.5m new motorcycles entered the market. If we extrapolate this to 6m new motorcycles per year, and assume that each new motorcycle replaces an older model (the fleet doesn't grow), it would take 9-years to reach 75% saturation.

   A regulatory approach can be seen as not imposing a direct cost on the Administration, however the national benefit from reduced casualties is dependent on the adoption of the technology. Therefore the road casualty cost reductions is realised equally slowly and remains as cost to the Administration.

   On a social equality measure, Regulation can be discriminatory. The early benefits of the new technology fall to those that are most able to afford to purchase new vehicles. The consumers in this category may not be the young or inexperienced riders that are prominent in the road casualty statistics. These road users may have to wait for the benefit until the new technology passes into the used vehicle market.

The pace at which a desired outcome will be achieved is an important policy consideration. As mentioned, a regulatory approach can be slow to implement but able to ensure an outcome. Consumer programmes can accelerate demand (and uptake) but depend on the consumer being persuaded and/or the manufacturer not wanting to be slower that their competitors in providing the best available safety.

Whichever approach is pursued, it is necessary to ensure that the anti-lock braking systems that enter the market are able to provide the level of safety that is expected. This requires legal clarity on what an "anti-lock braking system" is. This is important to prevent devices being placed on the market that are described as anti-lock but are not capable of delivering the expected safety improvements.

The nature of anti-lock technology requires that a legal definition includes both descriptive and performance elements. United Nations (UN) Regulation No. 78 and UN Global Technical Regulation (GTR) No. 3, made under the auspices of the World Forum for Harmonization of Vehicle Regulations, provide such clarity. The UN Regulations define an anti-lock braking function as follows:

"Antilock Brake System (ABS)" means a system which senses wheel slip and automatically modulates the pressure producing the braking forces at the wheel(s) to limit the degree of wheel slip."

The Legal Process

Overview

Legislation is often considered as the first choice to effect change and, if pursued, is dependent on political decisions. These decisions may be taken on purely political grounds, or influenced by public opinion and by evidence. Whatever the driver for legislation, it will be subject to the prevailing national legislative procedure. The actual process to be followed will depend, in part, upon the structure of laws in the individual countries. It can be expected that regulations governing road vehicles will be of a low level within a regulatory structure. Often there will be high level (primary) legislation that provides general principles and confers powers for lower level (secondary) legislation to be made. The secondary legislation will provide the detailed requirements.

The law making process will vary from country to country and will dictate the speed at which regulatory change can be effected. The process can be considered as having two elements, administrative, where the proposals (including impact assessments) are developed, and political, where decisions are taken.

Typically, both the policy development and the political elements of making secondary legislation can take twelve-months to complete. If primary legislation is required before the secondary legislation is possible, the time frame will be extended. The following chart shows the policy development process in Malaysia.

National Legislatures can be described as either, unicameral, where the decision making process is conducted in a single legislative chamber, or bicameral, where legislative proposals are scrutinised by two legislative chambers. Among the ASEAN countries, both systems are used.

Figure 2 **Malaysian policy development model**

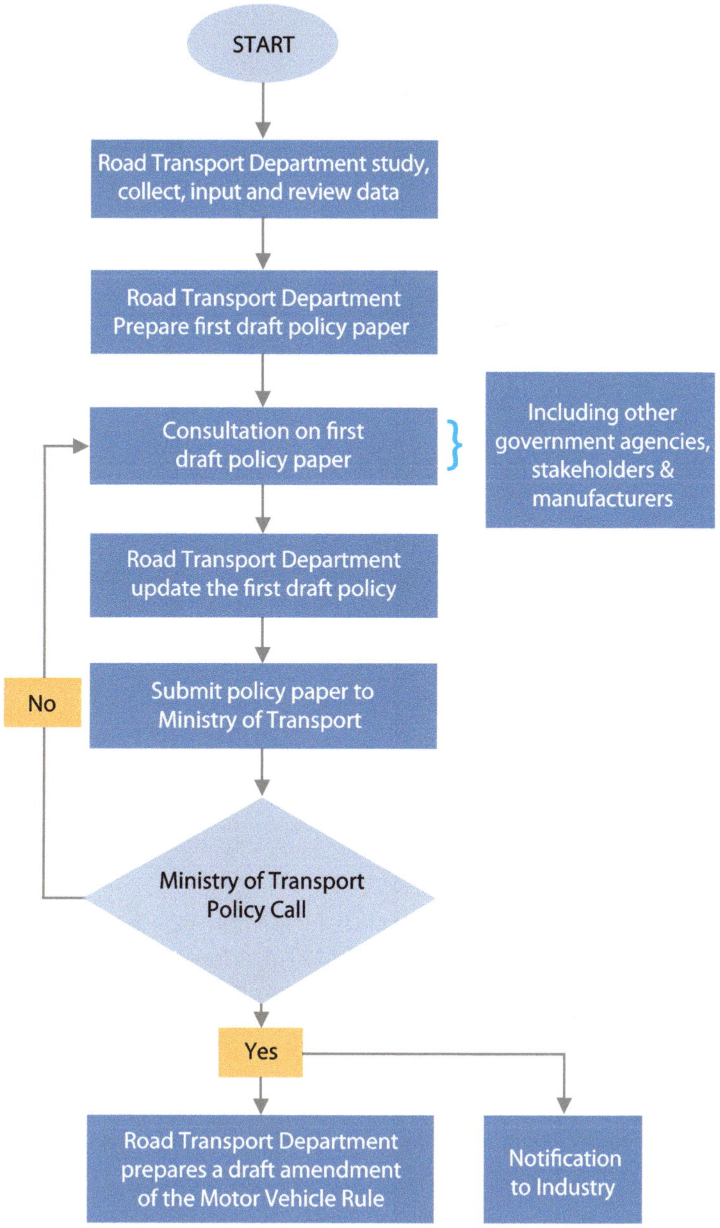

Source: Ts. Muhammad Firdhaus Ibrahim & Azzaharin bin Allias

Unicameral legislature

Viet Nam is one of four ASEAN countries that operates a unicameral system. Here, the Executive is responsible for proposals, and for developing the associated regulatory/policy impact assessment. Once complete, the proposals have to be verified by the National Assembly Agencies before the National Assembly Standing Committee can adopt the proposals into a draft legislative programme. This draft programme is subject to review and agreement by the National Assembly. The National Assembly meets twice a year with sessions in May and in October.

When content, the National Assembly will adopt proposals and make a "Law". Once adopted, a Law can provide for further legal instruments to be made to implement the objectives. If adopted (a Law is passed), it is likely that it will be necessary to produce a Decree and, possibly a Circular, to complete the implementation. Decrees are made by the government and may be supplemented by regulations containing the detail of the measures. Circulars are the responsibility of individual ministries and offer guidance on the administration of the Law or Decree.

Figure 3 **Unicameral Legislation procedure (Viet Nam)**

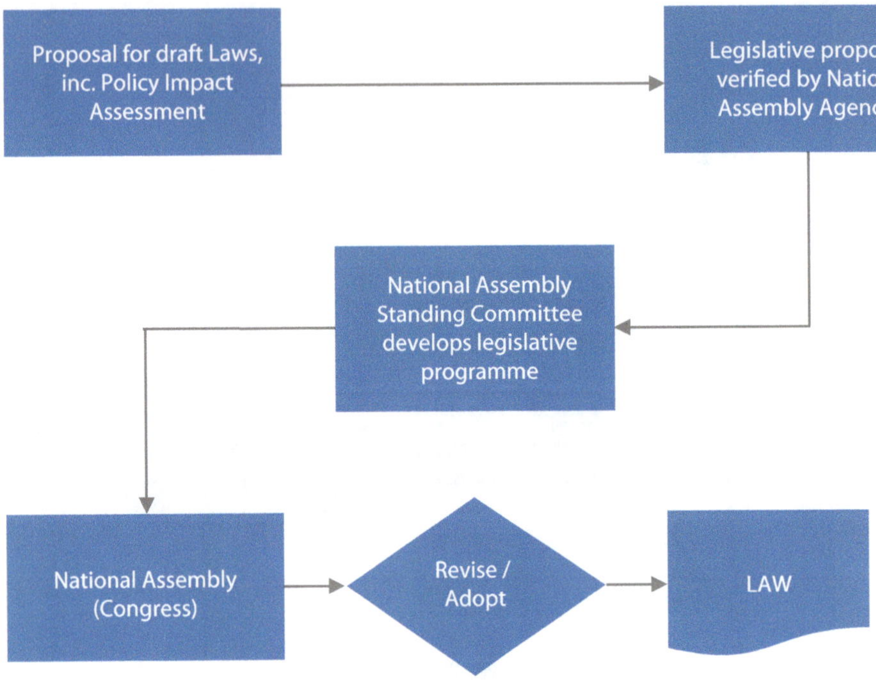

Source: Bernard Frost

Bicameral Legislature

Six ASEAN countries have a bicameral Legislature. In the Philippines, for example, high level (primary) legislation is presented to the political process as a Bill. Both the House of Representatives and the Senate receive the Bill and undertake a First Reading with the purpose of assigning the proposal to the relevant Committee for review. If the committee is content, the proposal then goes to a Committee on Rules who introduce the Bill into the legislative calendar.

Bills are subject to a debate at a Second Reading of the two chambers when amendments can be made. This is followed by a vote and, if favourable, the Bill is distributed for Third Reading in its final form. In the event of there being differences between the two chambers, a bicameral Conference Committee is formed to develop a resolution. Upon resolution, and the ratification of the Conference Committee report, the Bill is enrolled and sent for Presidential Action.

The Bill will become an Act if no there is action against it within 30 days. At this point the Act is published in the Official Gazette and takes effect 15 days later.

Figure 4 **Bicameral Legislation Procedure (Philippines)**

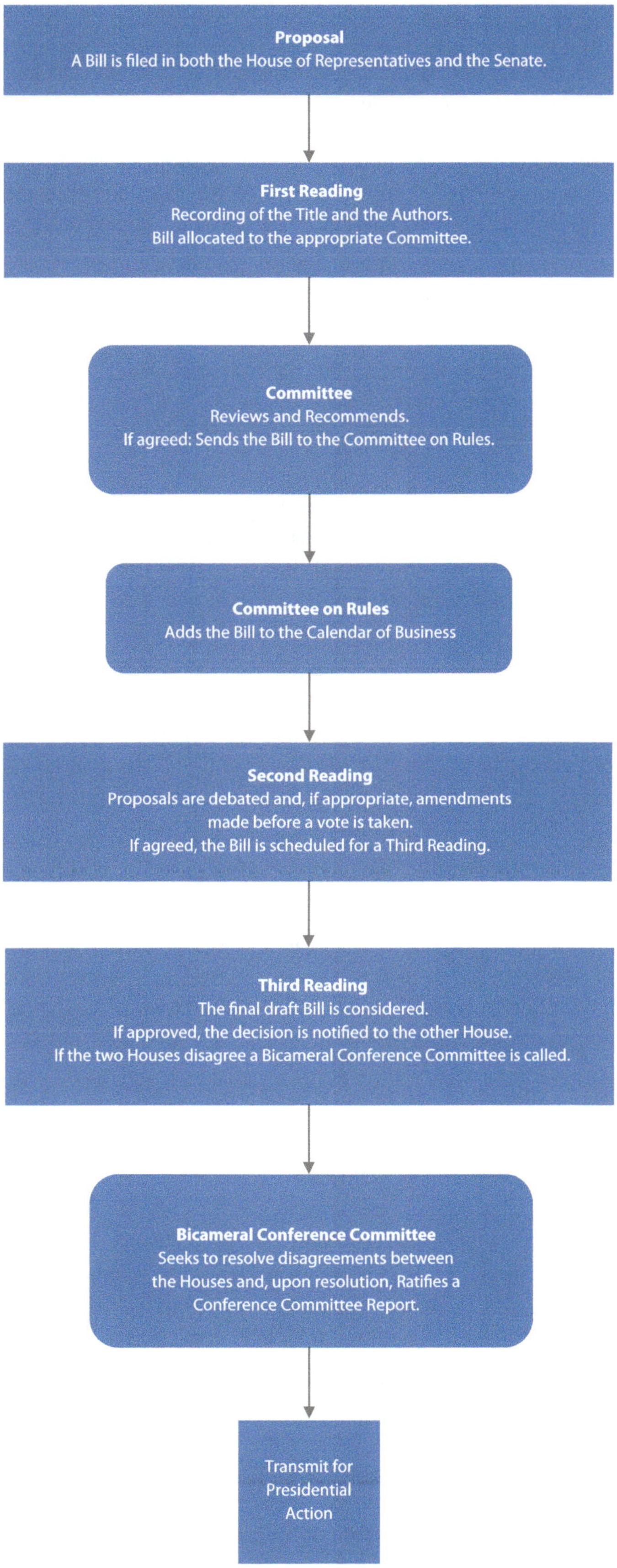

Source: Bernard Frost

The time taken to complete the bicameral process may vary between countries but is unlikely to be less than 12-months.

Motor Vehicle Legislation: Malaysia

The responsibility for regulations governing new policies of road vehicles lays with the Land Division of the Ministry of Transport (MOT). Enforcement of policies is the responsibility of the Road Transport Department (JPJ) which is an agency of the Land Division. Within the Road Transport Department, the Automotive Engineering Division is charged with implementing the specifications and technical requirements regarding safety regulations. The Division is supported with research by the Malaysia Institute of Road Safety Research (MIROS). However, any environmental policies, (e.g. exhaust emission and noise) on road vehicle are the responsibility of the Department of the Environment (DOE).

The specifications and technical requirements for motor vehicles are set out in the "Motor Vehicles (Construction and Use) Rules, 1959". This secondary legislation is made under powers provided by the "Road Transport Act 1987" (primary legislation).

Malaysia is a signatory to the UN 1958 Agreement and the UN 1998 Agreement. As of 2021, Malaysian legislation had gazetted 100 of the UN ECE Regulations made under the 1958 Agreement, including UN Regulation 78. There were a further 12 Malaysian Standards. The UN Regulations and Malaysian Standards are gazetted under the Motor Vehicles (Construction and Use) Rules, 1959.

New vehicles entering the Malaysian market are subject to Vehicle Type Approval. The type-approval procedure is governed by section 66(1) pp of the Road Transport Act and Rule 9A, of the Motor Vehicles (Construction & Uses) Rules. The Road Transport Department is the type approval Authority. Malaysia has committed to the mutual recognition of a type-approval certificate confirming compliance with a UN Regulation when issued by another type-approval Authority recognised under the 1958 Agreement.

The introduction of new technical requirements are subject to national consultation that includes a review of the costs, the benefits, and the impact on different sections of society. It also considers the impact on affected businesses (large and small) and, in particular, their ability to satisfy the new requirements within the proposed timeline. Typically this consultation lasts for 12-months. At the conclusion of the consultation process, it can take up to 12-months to complete the legal process and Gazette the amendment Rules.

Although UN Regulation 78 has been gazetted, type-approval is not dependent upon the fitment of anti-lock braking; consultation is required before the technology can be mandated. Once adopted, technical requirements are listed in the "Gazette of Motor Vehicles (Construction and Use) Rules, 1959".

The transitional provisions of UN Regulations are used as a main reference when developing proposals to implement a UN Regulation under type-approval. However, after the national consultation, the Road Transport Department will, in general, require that 2-years after listing in the Gazette, new types of vehicle entering the market must comply with the latest requirements. Again, in general, all newly registered vehicles (in-production model/ existing model) have to comply after a further 2-years.

In May 2023, Malaysia announced that it will, from 1 January 2025, mandate anti-lock braking on new model motorcycles (new vehicle type) with a capacity of 150 cm^3/13kW or above. At this time, it is not intended that this requirement will apply to all new motorcycles; existing types (in-production models) that have already been type-approved, and not equipped with anti-lock braking systems, will continue to be permitted for sale indefinitely.

Motor Vehicle Legislation: Viet Nam

In Viet Nam the highest level of legislation is a Law and is the preserve of the National Assembly (Congress). An example would be Law No. 23/2008/QH12 dated November 28, 2008 of the National Assembly on Road Traffic, prescribing measures for road traffic rules; road infrastructure facilities, vehicles in traffic and road users, road transportation and state management of road traffic.

Under the Law, the government can issue a Decree to regulate the powers available to ministries and to agencies. For example, Decree No. 51/2008/ ND-CP dated April 22, 2008 of the Government regulating the functions, tasks, powers, and organizational structure of the Ministry of Transport. With the powers given by the Decree, a Minister can issue a Circular that provides detailed and specific guidance on the general provisions of, for example, a Decree. As an example, Circular No. 45/2012/TT-BGTVT provides for the inspection of the technical quality, safety, and environment protection in the manufacture and assembly of motorcycles.

Lastly, a State Agency can issue Regulations setting out technical specifications and mandatory compliance requirements for specific subjects. The Regulation may refer to a Standard. A Standard can be published by a State Agency, by a social or professional organisation or by an economic organisation. While not a legal document, when referred to by a legal instrument a Standard assumes legal status for the purpose of that legal document. Regulation QCVN 14:2015/BGTVT contains the technical provisions for motorcycles, including their braking systems (but with no reference to anti-lock braking).

Following Viet Nam's accession to the UN 1958 Agreement, the government enacted Decree 60/2023 recognising that Agreement in respect to passenger cars. Viet Nam is considering how to address their accession to the UN 1958 Agreement in their national legal framework and, in particular, the implications of a type-approval system as provided for by that Agreement. Separately, Viet Nam has entered into a Free Trade Agreement with the European Union and it is expected that type-approvals issued by foreign type-approval authorities, including those for motorcycles, will be accepted soon.

To make ABS mandatory, it will be necessary for Viet Nam to make amendments to the Traffic Law and, at the time of writing, officials were still in the research phase of the impacts of anti-lock braking on the reduction of road accidents and the social and economic costs. If it is decided to propose amendments to the Traffic Law, other government agencies and the public (manufacturers associations, road safety organizations, etc.) are consulted can make suggestions for the content of the Law. After a period of time, if there is no objection, these amendments will be presented to the National Assembly (Congress) to vote on.

Finally, there will be new Decrees and Circulars enacted under these new amendments. The period of transition to comply with the new requirements may be established in the Law, Decree or Circular.

Motor Vehicle Legislation: The Philippines

Proposals for legislation are generally in response to a legislative programme announced by the President of the Philippines. It is normal for Executive Departments and Agencies prepare draft legislation and to submit it to both the House of Representatives and the Senate for consideration. However, business, and religious groups, consumers, trade associations, etc. may also propose legislation and a member of the Senate may sponsor the Bill for consideration.

The responsibility for technical standards affecting whole motor vehicles and systems rests with the Philippines' Department of Transportation (DOTr). It is understood that DOTr has not issued detailed technical regulations for new vehicles that enter the country, but motor vehicle type approval is under discussion following the accession of the Philippines to the UN 1958 Agreement. The Department of Trade and Industry - Bureau of Philippine Standards (DTI-BPS), has adopted the specifications and test methods of UN Regulation 78 as Philippine National Standard (PNS)/UN ECE 78:2006. However, PNS/standards are voluntary and are not referenced in technical regulations; compliance is neither mandatory nor checked.

At this time, no decisions have been taken within the Philippines with regard to making ABS a mandatory requirement on motorcycles, and the legislative procedure is not clear if such a decision was taken. It appears that, at this time, there are limited regulatory requirements for braking systems beyond an annual registration check that has parallels with a roadworthiness inspection. However, the Philippines has taken steps to improve the environmental performance of road vehicles by requiring compliance with UN Regulation 49.04 (heavy-duty vehicles) or Regulation 83.05B (light-duty vehicles). These provisions were made by "Implementing Rules"[10] made under the Philippine Clean Air Act 1999. Although not confirmed, it is likely that primary legislation would be necessary before any specific Implementing Rules on motorcycle braking (including anti-lock braking) could be made.

[10] *Department of Environment and Natural Resources Administrative Order, D.A.O 2016-23*

Implementation of UN Regulations

The UN regulations provide a flexible approach for national/regional Administrations to apply technical standards. National legislation may apply the technical requirements as a function of permitting the registration and sale of new vehicles into their territory and also legislate to ensure that those requirements continued to be fulfilled throughout the vehicle's life.

It is important to recognise that neither UN Regulation 78, nor UN GTR No.3., require anti-lock braking systems. Both instruments provide the methodology to assess the technology, "if it is fitted". Even if national legislation requires compliance with UN Regulation 78, it will be necessary for Administrations to be clear in that legislation that a motorcycle must be equipped with an anti-lock braking system that complies with the UN Regulation.

The technical standards for anti-lock braking in the Regulation are almost completely separate from the other requirements of the braking system. The only requirement tying the two together is for the motorcycle to meet at least the braking requirements for the rear brake if the anti-lock braking system experiences an electrical failure. While UN Regulation 78 is widely accepted as the standard for motorcycle braking, and it is recommended that it is included in national legislation, it would be possible, in national legislation, to refer to only the anti-lock element of UN Regulation 78 and retain existing national standards for the braking system itself.

The UN regulations provide the opportunity for Administrations to benefit from the technical requirements while following their own legislative approach. It is understood that there may be differences in the way that individual ASEAN countries wish to implement anti-lock braking requirements. These differences are likely to be associated with a legal definition that discriminates those vehicles affected by the legislation from those that are exempt.

Classification of Vehicles

The policy assessment associated with requiring anti-lock braking systems on motorcycles will identify whether it is desirable to require such technology on all motorcycles or only to a selected group. It may be that the benefit is associated with a distinct subset of those categories and that the cost of requiring anti-lock braking on the wider group is disproportionate to the benefit that may accrue.

Under the 1958 Agreement, motorcycles are classed as "Category L" vehicles and are defined by seven distinct sub-categories. The first discriminator of categories considers the number of wheels, two, three, or four. The second discriminator of categories considers the size or performance of the engine. The definitions are contained within the Consolidated Resolution (R.E.3)[11].

Similar definitions are used for UN GTRs, made under the 1998 Agreement where motorcycles are classified as Category 3 vehicles. These are contained in Special Resolution No. 1[12]. Here there are only five sub-categories as four-wheeled motorcycles (quad-bikes) are not included under the category 3 listing.

Table 1 **Motorcycle definitions according to Consolidated Resolution 3**

Two-Wheeled Vehicles	Three-wheeled Vehicles	Four-wheeled Vehicles
"Category L1": A two-wheeled vehicle with an engine cylinder capacity in the case of a thermic engine not exceeding 50 cm^3 and whatever the means of propulsion a maximum design speed not exceeding 50 km/h. In the case the vehicle is equipped with a "twinned wheels-" configuration, the full vehicle structure or part of the vehicle structure shall tilt when turning.	"Category L2": A three-wheeled vehicle of any wheel arrangement with an engine cylinder capacity in the case of a thermic engine not exceeding 50 cm^3 and whatever the means of propulsion a maximum design speed not exceeding 50 km/h.	"Category L6": A vehicle with four wheels whose unladen mass is not more than 350 kg, not including the mass of the batteries in case of electric vehicles, whose maximum design speed is not more than 45 km/h, and whose engine cylinder capacity does not exceed 50 cm^3 for spark (positive) ignition engines, or whose maximum net power output does not exceed 4 kW in the case of other internal combustion engines, or whose maximum continuous rated power does not exceed 4 kW in the case of electric engines.

[11] Classification and definition of Vehicles | 1958 Agreement

[12] Classification and Definition of Vehicles | 1998 Agreement

Two-Wheeled Vehicles	Three-wheeled Vehicles	Four-wheeled Vehicles
"Category L3": A two-wheeled vehicle with an engine cylinder capacity in the case of a thermic engine exceeding 50 cm³ or whatever the means of propulsion a maximum design speed exceeding 50 km/h. In the case the vehicle is equipped with a "twinned wheels-" configuration, the full vehicle structure or part of the vehicle structure shall tilt when turning.	"Category L4": A vehicle with three wheels asymmetrically arranged in relation to the longitudinal median plane with an engine cylinder capacity in the case of a thermic engine exceeding 50 cm³ or whatever the means of propulsion a maximum design speed exceeding 50 km/h (motor cycles with sidecars).	"Category L7": A vehicle with four wheels, other than that classified for the category L6, whose unladen mass is not more than 400 kg (550 kg for vehicles intended for carrying goods), not including the mass of batteries in the case of electric vehicles and whose maximum continuous rated power does not exceed 15 kW.
	"Category L5": A vehicle with three wheels symmetrically arranged in relation to the longitudinal median plane with an engine cylinder capacity in the case of a thermic engine exceeding 50 cm³ or whatever the means of propulsion a maximum design speed exceeding 50 km/h.	

Source: UNECE

While UN Regulations reference these categories, there is an opportunity for an Administration to introduce further discrimination into its legislation. For example, in the European Union (EU) vehicles of Category L3 are further defined by sub categories A1, A2 and A3. A sub-classification can be helpful to target specific vehicle groups that are considered to benefit most from the legislative measures.

Table 2 **European Union motorcycle sub-category definitions**

Sub-category L3e-A1	Low-performance motorcycle	(1) Engine capacity ≤ 125 cm³ and (2) Maximum continuous rated or net power ≤ 11 kW and (3) Power/weight ratio ≤ 0.1 kW/kg.
Sub-category L3e-A2	Medium-performance motorcycle	Maximum continuous rated or net power ≤ 35 kW and Power/weight ratio ≤ 0.2 kW/kg and Not derived from a vehicle equipped with an engine of more than double its power and L3e vehicle that cannot be classified under supplemental sub-classification criteria (1), (2) and (3) of an L3e-A1 vehicle.
Sub-category L3e-A3	High-performance motorcycle	Any other L3e vehicle that cannot be classified according to the classification criteria of an L3e-A1 or L3e-A2 vehicle.

Source: EU Regulation No 168/2013

The EU classification of "low-performance", "medium-performance", and "high-performance" motorcycle, illustrate how specific criteria can be employed to discriminate within a UN vehicle category. In these examples "power" and "power/weight" ratios have been used.

These classifications also recognise that, in addition to motorcycles with a thermic (internal combustion) engine, the use of electrically powered vehicles is growing. To protect the policy ambition for anti-lock braking systems national initiatives should consider similar provisions.

Some ASEAN countries have indicated that they are considering requiring anti-lock braking systems on motorcycles with engine capacities exceeding a particular size, 125 cm³ or 150 cm³ for example. Some road safety advocates point to the effect of speed on injury outcomes and that a speed discriminator should also be used when mandating the technology.

Legal Framework

Administrations will follow their own national model for legislating technical standards for road vehicles. Neither the 1958 Agreement nor the 1998 Agreement require an Administration to adopt a new or different approach.

Depending upon national convention, legislation can reference existing standards or, particularly for road vehicles, refer to regulations made under the World Forum for Harmonization of Vehicle Regulations. In some cases, it may be necessary to translate these texts into the national language and replicate it in national law. The availability of the text in the national language may influence that decision; all UN Regulations are produced in English, French and Russian.

The method by which the UN Regulations are implemented is also flexible. Administrations can choose:

1. Type Approval based upon the 1958 Agreement and the implementation of Mutual Recognition. This option would be used only by Contracting Parties to the 1958 Agreement who are also implementing UN Regulation 78 in their domestic legislation.
2. Type Approval based upon a national/regional programme that adopts the construction, fitting and performance requirements of UN Regulation 78 or UN GTR No.3, but with its own procedures for assessment and compliance checks.
3. Self-certification, using the construction, fitting and performance requirements of either UN Regulation 78 or UN GTR No.3.

Options 2 and 3 may be used by any Administration. If the Administration is a Contracting Party to the 1958 Agreement, it would have to advise the UN Secretary General that they are not implementing the Regulation fully (the 1958 Agreement requires such notification).

It should be noted that the structure of the UN Regulations is such that certain performance requirements apply to only some vehicle categories. For example, within UN Regulation 78, the high speed, and the heat fade tests do not apply to vehicles of Category L1 and L2. If the Administration is not already using the UN vehicle classifications, but intends to base its domestic legislation on UN Regulations, it should consider whether their domestic definitions ensure that appropriate tests are implemented.

Type Approval

The term "type-approval" indicates an administrative procedure by which an Authority declares, after carrying out the required verifications, that a type of vehicle, equipment or part, submitted by the manufacturer, conforms to the requirements of a given regulation. Generally, an Administration will accept that if a product is type-approved then all subsequent production will be of the same standard. The type-approval procedure is likely to include an assessment of the manufacturer's conformity of production procedures and is also likely to include routine inspections of the conformity of production records. The UN 1958 Agreement includes detailed provisions for both the technical and the administrative elements.

Administrations accept evidence of a type-approval as proof of compliance with national regulations at the time the vehicle is first registered. In addition to reviewing conformity of production, the Administration may also conduct market surveillance activities. The penalties for non-compliance, are not provided by the UN Agreements or the Regulations, and are usually set out in the national legislation.

Self-certification

Self-certification is an alternative method of regulating the construction, fitting and performance requirements for a vehicles, equipment or parts. It refers to a system whereby the national/regional legislation sets out those requirements, and the manufacturer is able to bring their product to the market on the basis that they themselves are satisfied that their design and production complies with those requirements.

National Administrations applying self-certification (including when implementing GTRs) will establish their own national or regional procedures for ensuring that products that are brought to their market comply with the legal requirements. Those procedures may not include an assessment of a product before it enters the market but, for example, include market surveillance procedures whereby production samples are assessed and, in the event of a non-compliance, legal action may be taken against the manufacturer. Again, penalties for non-compliance will be established in local legislation.

If an Administration choses to pursue a non-regulatory approach, it can still cite UN Regulations as a reference for any performance based policy. Depending on their policy, they may also need to distinguish those motorcycles that are required to be equipped and those that are not.

Options for adopting UN Regulations into national legislation

The previous section indicated three approaches for implementing the UN Regulations. These options are reviewed below.

Contracting Party or non-Contracting Party to the 1958 Agreement

The use of UN Regulations is not limited only to those Administrations that have signed the 1958 (or 1998) Agreement. The regulations are published and their use is not restricted in anyway.

By signing the 1958 Agreement (becoming a Contracting Party) and by implementing a particular UN Regulation, an Administration benefits in two ways. Firstly, it can influence the content of that Regulation by bringing forward proposals for amendments. Secondly, it has access to data provided by other Contracting Parties about the vehicles that have been type-approved to that regulation; this will include type approval detail and any conformity of production activities.

Type-Approval based upon the 1958 Agreement

A Contracting Party to the 1958 Agreement can use the full potential of that Agreement to ensure that products entering their market satisfy essential performance criteria. Unless a Contracting Party has indicated to the UN Secretary General that they will not implement a particular UN Regulation, then compliance with that UN Regulation has to be recognised as meeting national requirements. This "mutual recognition" of an approval, issued by a different Contracting Party, is an essential part of the 1958 Agreement and a corner stone to type-approval.

When an Authority issues a type-approval it is required to notify the other Contracting Parties that it has done so. The detail of what is required of the notification is included as an annex to each UN Regulation – Annex 1 in the case of UN Regulation 78 (Details of Annex 1 are replicated in Appendix 1 to this paper).

Details about a vehicle, equipment or part that has been type-approved are communicated to the other Contracting Parties, either by written notification or by inclusion on a central database -Database for the Exchange of Type Approval documentation (DETA). Contracting Parties to the 1958 Agreement can access DETA in respect to each of the UN Regulations to which they are signatories. Contracting Parties have to register and nominate a Focal Point for access to the database.

The type-approval notification provides the information necessary to identify the vehicle type and to establish the essential elements of its construction. In the case of UN Regulation 78, this information permits an Administration to identify those motorcycles that are equipped with an anti-lock braking system; this is not identified by the type approval number.

UN type-approval also requires a product to carry an indelible mark, in an accessible position, that indicates the approval. Figure 5 shows the arrangement of that mark:

Figure 5 **Arrangement of e-Mark**

Source: UNECE & Bernard Frost

The use of the capital letter "E", enclosed within a circle, indicates a type-approval to a UN regulation. The number next to the letter "E" identifies the Approval Authority that issued the approval. Each Contracting Party has its own unique identification number – in the example above the number "11" (E11) indicates that the approval was issued by the approval authority of the United Kingdom. The character sequence 78R-05 indicates that the approval has been granted in accordance with UN Regulation 78, as amended by the 05 series of amendments. Finally the last four digits provide the unique approval number for that approval.

To support mutual recognition, the 1958 Agreement requires that the production of vehicles, equipment or parts that have been type-approved is subject to regular manufacturing checks by the manufacturer. These checks are to ensure that the production versions of the type-approved product satisfy the same construction, fitting and performance requirements as the approved type. The manufacturer must have in place recognised quality procedures and processes for these "Conformity of Production" (CoP) checks. These quality procedures must include systems to record the CoP checks, including any production anomalies and the remedial measures that were undertaken. The Approval Authority that issued the approval is also required to conduct regular audits of the Conformity of Production to ensure continued compliance. The approval authority is obliged to inform the other Contracting Parties of any irregularities that are found and the measures taken to address them. Non-compliance can lead to recalls and, ultimately, to the type-approval being withdrawn.

As mentioned, to ensure that anti-lock braking systems are fitted to a motorcycle an Administration will have to make clear that compliance with UN Regulation 78 includes compliance with the relevant technical requirements for anti-lock braking systems (Annex 3). If the Administration chooses to use a regulatory approach the structure of the text might be framed as follows:

> *"Motorcycles first registered on or after [insert date], shall be type approved in accordance with UN Regulation 78, as amended by the 05 series of amendments (or later). For two-wheel motorcycles [having an engine capacity of 125 cm^3, or a maximum continuous rated or net power greater than 11 kW], the approval shall include compliance with the requirements of Annex 3, Section 9, of that UN Regulation."*

(This text is indicative, as it is the responsibility of lawyers to draft legislation)

Type-Approval based upon a national/regional programme

As an alternative to using UN type-approval, an Administration may employ a national type-approval scheme. This approach could be employed by an Administration that is not Contracting Party to the 1958 Agreement, or by a Contracting Party that, for whatever reason, does not wish to implement in full the requirements of the Regulation.

If this approach is adopted, the Administration could still take advantage of the technical provisions of the UN Regulations and require compliance with the construction, fitting and performance requirements of either UN Regulation 78 or UN GTR No.3.

The Administration could choose to recognise a UN type-approval as equivalent to their own national approval scheme but, even if it is a Contracting Party to the 1958 Agreement, it would not benefit from the communication of approval data associated with full implementation.

The Administration would be able to establish its own procedures for issuing approvals and for production conformity.

Lawyers may be able to use similar language to implement the requirements into national law as was indicated above. In this case of course type-approval would mean according to the national scheme.

Self-certification

In this approach the national legislation could reference the technical requirements of UN Regulation 78, or UN GTR No.3, and therefore capture the benefits of the UN Regulations in respect of safety performance. By referring to the UN technical requirements the Administration also provides their consumers with the benefit of global products that are likely to be built to these standards.

The manufacturer would be responsible for their interpretation of the technical requirements and that, using the procedures detailed in the relevant UN documents, the motorcycle complied.

The Administration will establish its own procedures for new vehicle registrations, perhaps substituting the evidence of type-approval with a manufacturer's declaration. The Administration will also have its own systems of market surveillance and enforcement to ensure that the products entering their market are compliant. Regulatory text could substitute the reference to type-approval in the earlier suggested text with a reference to the "construction, fitting and performance requirements":

> *"Motorcycles first registered on or after [insert date], shall comply with the construction, fitting and performance requirements of [UN Regulation 78, as amended by the 05 series of amendments (or later) / UN GTR No.3.including Amendments 1, 2 and 3], For two-wheel motorcycles [having an engine capacity of 125 cm^3 or a maximum continuous rated or net power greater than 11 kW] the requirements of Annex 3, Section 9, of that UN Regulation shall apply."*

(This text is indicative, as it is the responsibility of lawyers to draft legislation)

The examples refer to a "motorcycle". It should be noted that the structure of the UN Regulations is such that there are different requirements according to vehicle category. For example, the high speed, and the heat fade tests do not apply to vehicles of Category L1 and L2. If the Administration is not already using the UN vehicle classifications, it should consider whether their domestic definitions ensure that appropriate tests are implemented.

Obligations on Contracting Parties

A Contracting Party to the 1958 Agreement that intends to issue an approval to a particular UN Regulation must appoint an Approval Authority that has the responsibility for type-approval. The Approval Authority may undertake the verification testing itself, or appoint a third party as a "Technical Service" for this task; the appointed Technical Service need not be in the national territory of the Approval Authority.

Technical Services must be designated according to one or more of four categories of activities. The designation will depend upon their field of competence. The four categories are:

> Category A: Technical services which carry out the tests referred to in UN Regulations in their own facilities;
>
> Category B: Technical services which supervise the tests referred to in the UN Regulations, performed in the manufacturer's facilities or in the facilities of a third party;
>
> Category C: Technical services which assess and monitor on a regular basis the manufacturer's procedures for controlling conformity of production;
>
> Category D: Technical services which supervise or perform tests or inspections in the framework of the surveillance of conformity of production.

It is a responsibility of the Contracting Parties to ensure that Technical Services are properly assessed and to notify the UNECE secretariat of the name of the Technical Service through respective entry via the "/343 Application"[13] by the national focal point. Schedule 2 of the 1958 Agreement sets out the obligations for the assessment, designation and notification of technical services. It also sets out the standards which the technical services have to comply with. Details from Schedule 2 of the Agreement are replicated in Appendix 2.

[13] https://unece.org/data-sharing

The UN 1958 and 1998 Agreements

This paper has referred to UN Regulation 78 made under the UN 1958 Agreement. The technical standards for motorcycle braking are also present in UN GTR No.3, made under the UN 1998 Agreement. These two international agreements provide the legal framework under which technical Regulations for wheeled vehicles, equipment or parts, are developed. If an Administration chooses not to use type-approval as part of its strategy, it can achieve the same technical outcomes by referring to the construction, fitting and performance requirements of either UN Regulation 78 or UN GTR No. 3.

The 1958 Agreement has 60 Contracting Parties and has 167 UN Regulations within its orbit. The 1998 Agreement has 39 Contracting Parties and 24 Global Technical Regulations (GTR). Contracting Parties are members of the United Nations. While not each UN Regulation has a complementing GTR, in general, when a GTR is produced, or is amended, there will be parallel provisions within a UN Regulation.

The administration of the two agreements is the responsibility of the World Forum for Harmonization of Vehicle Regulations (Working Party 29) of the UNECE Inland Transport Committee. Working Party 29 has six subsidiary working parties that develop and amend the individual UN Regulations and GTRs. Contracting Parties, non-government organisations (NGOs) and industry, work together to ensure that the technical requirements are proportionate and relevant to the objective of delivering safer and/or more environmentally friendly vehicles. Only the Contracting Parties are able to decide on whether to adopt a new, or amend an existing, UN Regulation/GTR and the voting procedure is based upon a qualified majority.

The 1958 Agreement

The 1958 Agreement specifically provides for a process of type-approval. Any UN Regulation will include the administrative provisions for type-approval even where its technical requirements are mirrored in a GTR.

Type-approval, under the UNECE framework, is a procedure in which a manufacturer provides a representative sample of their product, the type, to the Type Approval Authority of a Contracting Party for assessment. The product is assessed against the construction, fitting and performance requirements of the particular regulation. The manufacturer is also assessed to determine whether it has the necessary quality control procedures in place to ensure that further production will be the same as the type presented for approval.

A core element of the 1958 Agreement is for the mutual recognition of a type-approval. Once a type-approval has been issued a Contracting Party to the 1958 Agreement is required to recognise that approval as satisfying their national requirements for that particular technology, e.g. a type- approval to the most recent version of UN Regulation 78 has to be accepted as meeting the national/regional requirements for motorcycle braking. Mutual recognition is an important element in reducing barriers to trade. However, being a Contracting Party does not require the Administration to mandate compliance with a UN Regulation, and it can set different national/regional standards, including recognition of an earlier version of the Regulation, as a minimum requirement.

Contracting Parties also have the option of not accepting a type-approval for one or more UN Regulations. If a Contracting Party wishes not to implement a particular UN Regulation, it can advise the UN Secretary General of its decision and the obligation under the 1958 Agreement for mutual recognition of an approval to that Regulation is annulled.

The 1998 Agreement

The 1998 Agreement, does not provide for type-approval and neither do GTRs made under it. There is therefore no obligation for a Contracting Party to accept a vehicle that has been demonstrated to comply with a GTR by another Contracting Party, and there is no obligation to communicate acceptance or rejection of a vehicle, equipment or part to another Contracting Party.

Contracting Parties applying GTRs will establish their own national or regional procedures for ensuring that products that are sold into their market comply with the legal requirements.

The application of a GTR by a Contracting Party can also assist in the removal of barriers to trade as it would recognise internationally agreed technical standards. Manufacturers have to determine which Contracting Parties are applying the GTR as a national standard. However, the manufacturer also has to ensure that they are able to provide the evidence of compliance to each market and this could vary from country to country. This approach is referred to as self-certification.

A Contracting Party that votes in favour of establishing a GTR is obliged to start a process of adopting it into its national law. The Contracting Party is required to advise the UN of its progress towards adoption. This obligation must be fulfilled on an annual basis and the status is reviewed by Working Party 29. As with the 1958 Agreement, a Contracting Party can choose to apply or to cease to apply a particular GTR but must notify the UN Secretary General of that decision.

It is worthy of note that a national Administration, particularly one that is not Contracting Party to an Agreement, can use regulations under the 1958 Agreement in much the same way as the 1998 Agreement. National law can simply refer to compliance with the construction, fitting and performance requirements of the particular regulation. As with the use of the 1998 Agreement, those Administrations would implement their own compliance and market surveillance measures.

Technical Elements of UN Regulation 78 / UN GTR No. 3.

In respect of motorcycle brakes UN Regulation 78 and GTR No. 3 offer equivalent technical provisions. Countries wishing to apply the technical provisions of UN regulations can therefore choose to refer to either UN Regulation 78 or UN GTR No. 3 Regulations contain detailed requirements for the construction, fitting and performance of braking systems for motorcycles. These requirements extend to all aspects of the braking system and are not limited to anti-lock braking.

These regulations provide for the assessment of six different aspect of the braking system with a series of dynamic tests are detailed to evaluate compliance with the performance requirements. These provisions are applicable as follows:

Table 3 **UN Regulation 78 test requirements**

	Dry Test	High Speed Test	Heat/ Fade Test	Wet Brake Test	Parking Brake Test	Anti-lock Brake Test
L1 (3-1)	X					
L2 (3-2)	X				X	
L3 (3-3)	X	X	X			
L4 (3-4)	X	X	X	Applicable to all categories unless the brakes are enclosed.	X	Applicable to all, if ABS is fitted
L5 (3-5)	X	X	X		X	
L6	X					
L7	X	X	X		X	

Source: UNECE

1. A Dry Test

This test is conducted in engine disconnected state, in the laden condition and, in the case of a vehicle with combined braking system or a split service brake system, also in the lightly loaded condition.

Each brake system (front and rear) is tested separately and with a limit set for the force to be applied to the control. Categories L1, L2 and L6 are tested at 40 km/h and those of Categories L3, L4, L5 and L7 at 60 km/h.

Performance requirements are listed for each brake and the vehicle must satisfy either a minimum stopping distance or a minimum mean fully developed deceleration.

Additionally vehicles of Categories L3, L4, L5 and L7, are tested in the lightly loaded condition at 100km/h (or 90% their maximum speed) using all of the service brakes. For this test, a minimum stopping distance is defined.

2. A High-Speed Test (not applicable to L1, L2 and L6)

This test applies to vehicles of Categories L3, L4, L5 and L7 that have a maximum speed greater than 125 km/h. The vehicle is tested in a lightly loaded condition, with the engine connected, and maximum control forces are prescribed.

Vehicles with a maximum speed greater than 125 km/h but less than 200 km/h are tested at 80% of their maximum speed. Vehicles capable of speeds greater than 200 km/h are tested at 160 km/h.

The vehicle must satisfy either a minimum stopping distance or a minimum mean fully developed deceleration.

3. A Wet Brake Test (not applicable to enclosed brakes)

In this test the performance of the brake when dry is compared with that when it is subject to a wetting process. It applies to all categories but does not apply to drum brakes (unless they have a ventilation or open inspection ports) or fully enclosed disc brakes.

This test is conducted in engine disconnected state, in the laden condition and, in the case of a vehicle with combined braking system or a split service brake system, also in the lightly loaded condition.

The test conditions are as for the dry stop test but with a control force that delivers a deceleration of between 2.5 and 3.0 m/s^2.

A comparison is made between the performance of the brake when it is dry and when it has been subjected to wetting. Measurements are taken at between 0.5 and 1.0 second during the braking event and over the entire braking event. For the first comparison the wet brake performance must not be less than 60% of the dry brake. For the full braking stop, the wet brake performance must not exceed 120% of the dry brake performance.

4. A Heat/Fade Test (not applicable to L1, L2 and L6)

This test applies to vehicles of Categories L3, L4, L5 and L7 and comprises three parts:

- A baseline test using the dry stop procedure with a single control,
- A series of 10 stops, with a prescribed interval between each one, to create heat in the brake,
- A hot brake stop, again using the procedure of the dry stop test.

The tests are conducted in the laden condition with the engine disconnected. The test speed is 60 km/h or 90% of the maximum speed, whichever is lower.

The hot brake must deliver a stopping distance that is proportionate to the baseline test or a mean fully developed deceleration that is not less than 60% of the baseline test.

5. A Parking Brake Test (not applicable to a two-wheeled motorcycle)

Applicable to vehicles of categories L2, L4, L5, and L7, this test is conducted in the laden condition with the engine disconnected. The vehicle is parked on an 18% gradient and the parking brake applied with the control force limited by the Regulation.

The parking brake must hold the vehicle stationary on the gradient for at least 5 minutes. The vehicle is tested being parked both uphill and downhill.

6. An anti-lock braking system test (only if the motorcycle is fitted with ABS)

a) A brake performance test and vehicle stability test on a high adhesion surface.

This test is conducted at 60 km/h or 90% of the maximum speed, whichever is the lower. Both the front and the rear wheels are braked simultaneously. If one wheel is not equipped with ABS the control force is managed to avoid wheel locking.

There must be no wheel locking, the vehicle must stay in a 2.5metre lane and a minimum stopping distance or mean fully developed deceleration must be achieved.

b) A brake performance test and vehicle stability test on a high adhesion surface.

This a repeat of the test on a high adhesion surface. Again there must be no wheel locking and the vehicle must stay within the test lane. However, here the friction value of the test surface is included in the calculation of the required stopping distance or mean fully developed deceleration.

c) Wheel lock checks on both a high and a low adhesion surface.

For the high friction surface the test speed is 80 km/h, on the low adhesion surface it is 60 km/h or, in either case, 80% of the maximum speed, whichever is lower.

Each brake is tested separately with the ABS fully cycling down to a speed of 10 km/h. In addition, if both wheels are controlled by the ABS, a further test is conducted with the ABS fully cycling on both wheels.

There must be no wheel locking, the vehicle must stay in the 2.5 metre test lane

d) Wheel lock check on a high to low adhesion surface transition.

For this test there must be an immediate change in the friction value of the test track within the test lane. At the point that the vehicle transitions from the high to low adhesion surface, it must be travelling at 50 km/h (or 50% of its maximum speed, whichever is lower) with the ABS fully cycling.

The brakes of each wheel are tested separately and, where both wheels are equipped with ABS, a further test is performed with the brakes actuated on both wheels.

There must be no wheel locking, the vehicle must stay in the 2.5 metre test lane

e) Wheel lock check on a low to high adhesion surface transition.

This test follows the procedures of the high to low transition test above but with the transition to a high adhesion surface.

In this case, in addition to there being no wheel lock, and the vehicle remaining in the test lane, it has to be demonstrated that the ABS has reacted to the availability of a higher friction surface by increasing the braking rate.

f) Performance test with an ABS electrical failure.

Lastly, the braking system is tested with a failure of the electrical element of the ABS. For this test the conditions used for Dry Stop test with a single control actuation.

The test speed will be 40 km/h for vehicles of category L1, L2 and L6, or 60 km/h for vehicles of category L3, L4, L5 or L7.

Again in both cases the speed may be at 90% of the vehicle's maximum if this is lower.

The vehicle must achieve the appropriate stopping distance, or mean fully developed deceleration, for secondary braking.

Conclusions

Casualty rates for motorcyclists are a subject for concern in all regions of the world but in low to middle income Regions, where their use makes powered transport accessible, the importance of risk mitigation should be paramount.

There is a large body of research that concludes that anti-lock braking systems, even when operating on only the front wheel of a motorcycle, can have a significant effect on loss of control and on collisions. It has been shown that anti-lock braking systems can help to shorten stopping distances especially on paved roads, and that, with minimal training/information, even inexperienced riders can achieve stopping distances close to those of experienced riders.

Anti-lock braking technology has improved in performance, reduced in size and in mass, and its cost has fallen. Some observers have suggested that a single channel system could be provided at a cost of around 100-150 Euros or potentially significantly less (depending on economy of scales). These factors have helped to make the technology suitable for even small, low value, motorcycles and, for example, in the European Union, anti-lock braking is now required on all new motorcycles with an engine capacity exceeding 125cm3.

In promoting the technology it is important that the consumer can have confidence that it will fulfil essential performance criteria.

Regulation 78 is concerned with motorcycle braking and includes a discrete set of requirements for anti-lock braking. The UN Regulations are developed under the framework of UN 1958 Agreement, which supports the mutual recognition by one country of a type-approval issued by another. This helps to reduce barriers to trade, but it can also help speed the transition to a new technology as manufacturers can provide a globally recognised product to new markets.

Use of the UN regulations is not restricted, and they can be adopted and adapted to suit the political and legal framework of any Administration.

If UN Regulation 78 (or UN Global Technical Regulation No. 3) is chosen as the technical standard for motorcycle braking, it has to be recognised that the national intervention needs to be specific regarding the requirement for anti-lock braking. While the construction and performance requirements for the technology are contained in both regulatory texts, the fitment is not mandatory, even for type-approval.

A political decision for intervention may require a comprehensive analysis of the impact of the measure. This analysis requires, as a minimum, a benefit to cost assessment but, more typically, will also require an understanding of the impact of the intervention on the population, often in terms of their social and economic background, on manufacturers and on businesses. Typically, this analysis is provided as part of a policy, or regulatory, impact assessment and forms part of a consultation with the affected parties.

Mandating Motorcycle ABS

Thailand for new models as of Jan. 2024 and will become mandatory for all (existing) models by Jan. 2026 (ABS: > 125ccm or 11 kW or 0.1 kW/kg; ABS/CBS: < 125ccm or 11 kW or 0.1 kW/kg)

Malaysia for new models as of Jan. 2025 (ABS: > 150ccm or 13kW for electric motorcycles)

Viet Nam for new models expected as of 2025 (ABS/CBS: >50 km/h)

Recommendations

The cornerstone of government policy is generally a strong evidence base. This evidence is commonly presented in the form of a Regulatory, or Policy, Impact Assessment. In the case of road safety initiatives this invariably requires a detailed understanding of the collision and casualty profile in the country. However, the WHO has identified that these casualty data are not easily determined in all countries. They also note that the casualty numbers in their own data often exceed those contained in official national data. This can be explained by differences in the way that loss of life is attributed to a road collision, for example, the number of days a casualty survives can determine whether their death is attributed to the collision.

The 2016 ASEAN Road Safety Strategy identified the adoption of anti-lock braking on motorcycles as a specific ambition. However, the delivery of the ambition, particularly if it is driven by mandatory requirements, can have resource implications, particularly for those countries where data are less complete. The use of Policy/Regulatory Impact Assessments is a widely accepted approach to policy development and often an essential element before there is political engagement. The work of the WHO in developing tools to overcome limited, or incomplete, data provides an opportunity to build a benefit to cost profile (an essential element of an impact assessment), based on a common approach, for the individual countries of the ASEAN Region.

The shared interests of the UNESCAP, UNECE, UNRSF, and WHO, in developing road safety interventions, suggest that this initiative should be able to be promoted under the UN Road Safety Collaboration (UNRSC); this should be explored. While not a complete impact assessment, such modelling could influence political priorities.

Recognising that the 2016 ASEAN Road Safety Strategy is a 10-year vision, it may be reasonable to think that a new, or renewed, strategy may be developed for 2026 and beyond. If that is the case, the presentation of a detailed report, using WHO methods, and setting out casualty and economic benefits of the mandatory uptake of motorcycle anti-lock braking, can only raise the profile of the initiative. Separately, individual governments could be approached with the detail of such an analysis which could act as a catalyst for change.

Any approach should also promote the merits of adopting the technical regulations developed by the UNECE World Forum for the Harmonization of Vehicle Regulations as part of an intervention. These Regulations (UN Regulation No. 78 or UN Global Technical Regulation No. 3) provide ready-made construction and performance requirements that have been widely adopted by the motorcycle manufacturing sector. They are suitable for adoption into any regulatory regime whether that requires type-approval (with or without mutual recognition of foreign approvals) or a system of self-certification or compliance.

Separately, there should be careful consideration of opportunities to promote a non-regulatory approach. Depending on the outcome of the benefit to cost assessment, this could be a targeted incentive scheme or simply a consumer awareness initiative. An incentive scheme does not necessarily mean that it renders a cost to the national administration, for example would the insurance industry receive a benefit and could they reflect the fitting of anti-lock systems in their premiums? Non-regulatory methods can make an impact on the complexion of the national fleet more quickly than a regulatory intervention.

Research has shown that motorcyclists benefit from guidance and instruction regarding the use of anti-lock braking. Consideration should be given to how to inform and educate a population of the benefits of the technology, particularly where the risk of wheel lock has resulted in habitual under-use of the front brake.

Appendix

Appendix 1: UN Regulation 78 – Annex 1

Annex 1[*]

Communication

(Maximum format: A4 (210 x 297 mm)

Issued by: Name of administration:

....

....

....

....

Concerning:[2] Approval granted

Approval extended

Approval refused

Approval withdrawn

Production definitively discontinued

of a type of vehicle of category L with regard to braking pursuant to UN Regulation No. 78

Approval No. .... Extension No. ....

1. Trade name or mark of the vehicle..

2. Vehicle type designation by the manufacturer..

3. Name and address of the manufacturer: ....

4. Name and address of the manufacturer's representative

(if applicable): ....

5. Summarized description: ....

5.1. Vehicle:

Category of vehicle: ....

Gross vehicle mass of vehicle: ....

5.2. Engine: ....

5.3. Transmission:

Number and ratios of gears: ....

Final drive ratio: ....

Tyre dimensions: ....

5.4. Braking system: ....

Make(s) and type(s) of linings: ....

Service brake(s) (front, rear, combined)[2]

Secondary braking, parking brake (if applicable)[2]

Other systems (anti-lock brakes, etc.)

6. Technical Service conducting approval tests: ....

7. Date of test report: ....

8. Number of test report: ....

9. Reason for extension of approval (if applicable): ....

10. Other remarks (if applicable), (right or left hand driven): ....

11. Place: ....

12. Date: ....

13. Signature: ....

14. Annexed the list of parts constituting the approval documents, which can be obtained on request, submitted to the Type Approval Authority which has delivered the approval.

* At the request of (an) applicant(s) for UN Regulation No. 90 approval, the information shall be provided by the Type Approval Authority, as contained in Appendix 1 to this annex. However, this information shall not be provided for purposes other than UN Regulation No. 90 approvals.

[1] Distinguishing number of the country which has granted/extended/refused/withdrawn approval (see approval provisions in the Regulation).

[2] Strike out what does not apply.

Annex 1 – Appendix

List of vehicle data for the purpose of UN Regulation No. 90 approvals

1. Description of the vehicle type: ....

1.1. Trade name or mark of the vehicle, if available: ....

1.2. Vehicle category: ....

1.3. Vehicle type according to UN Regulation No. 78 approval: ....

1.4. Models or trade names of vehicles constituting the vehicle type, if available:

....

1.5. Manufacturer's name and address: ....

2. Make and type of brake linings: ....

3. Minimum mass of vehicle: ....

3.1. Distribution of mass of each axle (maximum value): ....

4. Gross vehicle mass of vehicle: ....

4.1. Distribution of mass of each axle (maximum value): ....

5. Vmax .... km/h.

6. Tyre and wheel dimensions: ....

7. Configuration of the independent braking systems: ....

8. Specifications of brake valves (if applicable): ....

8.1. Adjustment specifications of the load sensing valve: ....

8.2. Setting of pressure valve: ....

9. Specification of brake: ....

9.1. Disc brake type (e.g. number of pistons with diameter(s), ventilated or solid disc):

....

9.2. Drum brake type (e.g. simplex, with piston size and drum dimensions): ....

10. Master cylinder type and size (if applicable): ....

Appendix 2: 1958 Agreement - Schedule 2

Schedule 2

Part one: Assessment, designation and notification of technical services

1. Designation of technical services

1.1. When an approval authority designates a technical service, the latter shall comply with the provisions of this Schedule.

1.2. The technical services shall carry out themselves, or supervise, the tests required for approval or inspections specified in UN Regulations, except where alternative procedures are specifically permitted. They may not conduct tests or inspections for which they have not been duly designated.

The performance of technical services and the quality of the tests and inspections they are carrying out shall ensure that the products for which UN type approval is requested are adequately verified for their compliance with the requirements of the applicable UN Regulations for which the technical services are designated.

1.3. The technical services shall be designated according to one or more of the four following categories of activities, depending on their field of competence:

(a) Category A: Technical services which carry out the tests referred to in UN Regulations in their own facilities;

(b) Category B: Technical services which supervise the tests referred to in the UN Regulations, performed in the manufacturer's facilities or in the facilities of a third party;

(c) Category C: Technical services which assess and monitor on a regular basis the manufacturer's procedures for controlling conformity of production;

(d) Category D: Technical services which supervise or perform tests or inspections in the framework of the surveillance of conformity of production.

1.4. Technical services shall demonstrate appropriate skills, specific technical knowledge and proven experience in the specific fields covered by the UN Regulations for which they are designated.

In addition, technical services shall comply with, but not necessarily be approved/accredited in accordance with, the standards listed in Part two of this Schedule which are relevant for the categories of activities for which they are designated.

The technical services shall ensure that they are free from any control and influence of interested parties which may adversely affect the impartiality and quality of the tests and inspections.

The technical services shall have access to the test facilities and measurement devices necessary to supervise or perform tests or inspections referred to in the UN Regulations for which the technical services are designated.

1.5. An approval authority may act as a technical service for one or more of the activities referred to in paragraph 1.3. Where an approval authority acting as a technical service has been appointed by national law of a Contracting Party and is financed by the latter, the provisions of this Schedule or equivalent rules to the provisions in paragraphs 1, 2 and 3.4 of this Schedule shall be complied with. The same applies for technical services that have been appointed by national law of a Contracting Party and are subject to financial and managerial control by the Government of that Contracting Party. The equivalent rules shall guarantee the same level of performance and independence.

1.6. Regardless of paragraph 3.3, a manufacturer or its representative acting on its behalf may be designated as a technical service for category A activities for only those UN Regulations which make provision for such a designation. In this case, and regardless of paragraph 1.4, such technical service shall be accredited in accordance with the standards referred to in paragraph 1 of Part two of this Schedule.

1.7. The entities referred to in paragraphs 1.5 and 1.6 shall comply with the provisions of paragraph 1.

2. Assessment of the skills of the technical services

2.1. The skills referred to in paragraph 1 shall be demonstrated in an assessment report established by a competent authority.[14] This may include a certificate of accreditation issued by an accreditation body.

2.2. The assessment referred to in paragraph 2.1 shall be conducted in accordance with the provisions of Part three of this Schedule.

The assessment report shall be reviewed after a maximum period of three years.

2.3. The assessment report shall be communicated to the UNECE secretariat and to the Contracting Parties upon request.

2.4. The approval authority which acts as a technical service shall demonstrate compliance with documentary evidence.

This includes an assessment which shall be conducted by auditors independent of the activity being assessed. Such auditors may be from within the same organization provided that they are independent of the personnel undertaking the assessed activity.

2.5. A manufacturer or its representative acting on their behalf, designated as the technical service, shall comply with the relevant provisions of paragraph 2.

3. Procedures for notification

3.1. Contracting Parties shall notify the UNECE secretariat of the name, the address including electronic address and the category of activities of each designated technical service. They shall also notify the UNECE secretariat any subsequent modifications thereof.

The notification act shall state for which UN Regulations the technical services have been designated.

3.2. A technical service may conduct the activities described in paragraph 1 for the purposes of UN type approval only if it has been notified to the UNECE secretariat.

3.3. The same technical service may be designated and notified by several Contracting Parties, irrespective of the category of activities which they conduct.

3.4. The UNECE secretariat shall publish the list and contact details of the approval authorities and technical services on its website.

Part two: Standards which the technical services, referred to in Part one of this Schedule, shall comply with

1. Activities in testing for UN type approval, to be carried out in accordance with UN Regulations

1.1. Category A (tests performed in own facilities):

ISO/IEC 17025:2005 on the general requirements for the competence of testing and calibration laboratories.

A technical service designated for category A activities may carry out or supervise the tests according to UN Regulations for which it has been designated, in the facilities of the manufacturer or of its representative.

1.2. Category B (supervising tests performed in the manufacturer's facilities or in the facilities of its representative):

ISO/IEC 17020:2012 on the general criteria for the operation of various types of bodies performing inspection.

Before performing or supervising any test in the facilities of a manufacturer or of its representative, the technical service shall verify that the test facilities and measurement devices comply with the appropriate requirements of paragraph 1.1.

2. Activities related to Conformity of Production

2.1. Category C (procedure for the initial assessment and surveillance audits of the manufacturer's quality management system):

ISO/IEC 17021:2015 on the requirements for bodies providing audit and certification of management systems.

2.2. Category D (inspection or testing of production samples or supervision thereof):

ISO/IEC 17020:2012 on the general criteria for the operation of various types of bodies performing inspection.

---

[14] "Competent authority" means either the approval or designated authority, or an appropriate accreditation body acting on their behalf respectively.

Part three: Procedure for the assessment of technical services

1. Purpose

1.1. This part of Schedule 2 establishes the conditions by which the assessment procedure of the technical services shall be conducted by the competent authority referred to in paragraph 2 of Part one of this Schedule.

1.2. These requirements shall apply, mutatis mutandis, to all technical services irrespective of their legal status (independent organization, manufacturer or approval authority acting as technical service).

2. Principles of assessing

Assessment shall be characterized by reliance on a number of principles:

(a) Independence which is the basis for the impartiality and objectivity of the conclusions;

(b) An evidence-based approach which guarantees reliable and reproducible conclusions.

Auditors shall show trust and integrity, and shall respect confidentiality and discretion. They shall report truthfully and accurately findings and conclusions.

3. Auditor skills

3.1. The assessments may only be conducted by auditors with the technical and administrative knowledge necessary for such purposes.

3.2. The auditors shall have been trained specifically for assessment activities. In addition, they shall have the specific knowledge of the technical area in which the technical service will exercise its activities.

3.3. Without prejudice to the provisions of paragraphs 3.1 and 3.2 above, the assessment referred to in paragraph 2.5 of Part one of this Schedule shall be conducted by auditors independent of the activities for which the assessment is conducted.

4. Application for designation

4.1. A duly authorized representative of the applicant technical service shall make a formal application to the competent authority. The application shall include as a minimum the following:

(a) General features of the technical service, including corporate entity, name, addresses, legal status and human and technical resources;

(b) A detailed description including curriculum vitae of the personnel in charge of testing and/or inspections and of the managerial staff as evidenced by the skills both educational and professional;

(c) In addition to the above, technical services which use virtual testing methods shall provide evidence of their ability to work in a Computer-Aided-x environment;

(d) General information concerning the technical service, such as its activities, its relationship in a larger corporate entity, if any, and addresses of all its physical location(s) to be covered by the scope of designation;

(e) An agreement to fulfil the requirements for designation and the other obligations of the technical service as applicable in the relevant UN Regulations for which it is designated;

(f) A description of the conformity assessment services that the technical service undertakes in the framework of the applicable UN Regulations and a list of the UN Regulations for which the technical service applies for designation, including its limits of capability where applicable;

(g) A copy of the quality assurance manual or comparable operational rules of the technical service.

4.2. The competent authority shall review for adequacy the information supplied by the technical service.

4.3. The technical service shall notify the approval authority of any modifications to the information provided in accordance with paragraph 4.1.

5. Resource review

The competent authority shall review its ability to carry out the assessment of the technical service, in terms of its own policy, its competence and the availability of suitable auditors and experts.

6. Subcontracting the assessment

6.1. The competent authority may subcontract parts of the assessment to another designated authority or ask for support from technical experts provided by other competent authorities. The subcontractors and experts shall be accepted by the applicant technical service.

6.2. The competent authority shall take into account accreditation certificates with adequate scope, in order to complete its global assessment of the technical service.

7. Preparation for assessment

7.1. The competent authority shall formally appoint an assessment team. The former shall ensure that the expertise brought to each assignment is appropriate. In particular, the team as a whole:

(a) Shall have appropriate knowledge of the specific scope for which designation is sought; and

(b) Shall have sufficient understanding to reliably assess the competence of the technical service which operates within its scope of designation.

7.2. The competent authority shall clearly define the assignment given to the assessment team. The task of the assessment team is to review the documents collected from the applicant technical service and to conduct the on-site assessment.

7.3. The competent authority shall agree, together with the technical service and the assigned assessment team, on the date and timetable for the assessment. However, it remains the responsibility of the competent authority to pursue a date that is in accordance with the surveillance and reassessment plan.

7.4. The competent authority shall ensure that the assessment team is provided with the appropriate criteria documents, previous assessment records, and the relevant documents and records of the technical service.

8. On-site assessment

The assessment team shall conduct the assessment of the technical service on the premises of the technical service from which one or more key activities are performed and, where relevant, shall perform eyewitness assessment at other selected locations where the technical service operates.

9. Analysis of findings and assessment report

9.1. The assessment team shall analyse all relevant information and evidence from the document and record review and the on-site assessment. This analysis shall be sufficient to allow the team to determine the extent of competence and conformity of the technical service with the requirements for designation.

9.2. The competent authority's reporting procedures shall ensure that the following requirements are fulfilled.

9.2.1. A meeting shall take place between the assessment team and the technical service prior to leaving the site. At this meeting, the assessment team shall provide a written and/or oral report of its findings obtained from the analysis. The technical service shall have the opportunity to ask questions about the findings, including non-conformities, if any, and their basis.

9.2.2. A written report on the outcome of the assessment shall be promptly brought to the attention of the technical service. This assessment report shall contain comments on competence and conformity, and shall identify non-conformities, if any, that need to be resolved in order to conform to all of the requirements for designation.

9.2.3. The technical service shall be invited to respond to the assessment report and to describe the specific actions taken or planned to be taken, within a defined time, to resolve any identified non-conformities.

9.3. The competent authority shall ensure that the responses of the technical service are sufficient and effective to resolve non-conformities. If the technical service responses are found to be insufficient, further information shall be requested. Additionally, evidence of effective implementation of actions taken may be requested, or a follow-up assessment may be carried out to verify effective implementation of corrective actions.

9.4. The assessment report shall include, as a minimum:

(a) The unique identification of the technical service;

(b) The date(s) of the on-site assessment;

(c) The name(s) of the auditor(s) and/or experts involved in the assessment;

(d) The unique identification of all premises assessed;

(e) The proposed scope of designation that was assessed;

(f) A statement on the adequacy of the internal organization and procedures adopted by the technical service supporting its competence, as determined through its fulfilment of the requirements for designation;

(g) The information on resolving all non-conformities;

(h) A recommendation on whether the applicant should be designated or confirmed as technical service and, if so, the scope of designation.

10. Granting/confirming a designation

10.1. The approval authority shall, without undue delay, make the decision on whether to grant, confirm or extend designation on the basis of the report(s) and any other relevant information.

10.2. The approval authority shall provide a certificate to the technical service. This certificate shall identify the following:

(a) The identity and logo of the approval authority;

(b) The unique identity of the designated technical service;

(c) The effective date of granting of designation and the expiry date;

(d) A brief indication of, or a reference to, the scope of designation (applicable UN Regulations or parts thereof);

(e) A statement of conformity and a reference to this Schedule.

11. Reassessment and surveillance

11.1. Reassessment is similar to an initial assessment except that experience gained during previous assessments shall be taken into account. Surveillance on-site assessments are less extensive than reassessments.

11.2. The competent authority shall design its plan for reassessment and surveillance of each designated technical service, so that representative samples of the scope of designation are assessed on a regular basis.

The interval between on-site assessments, whether reassessment or surveillance, depends on the proven stability of the technical service.

11.3. When, during surveillance or reassessments, non-conformities are identified, the competent authority shall define strict time limits for the implementation of corrective actions.

11.4. When the corrective or improvement actions have not been taken within the agreed timeframe, or are not deemed to be sufficient, the competent authority shall adopt appropriate measures such as, conducting further assessment, suspending/withdrawing the designation for one or more of the activities for which the technical service has been designated.

11.5. When the competent authority decides to suspend or withdraw the designation of a technical service, it shall inform the latter by registered mail, and shall inform the UNECE secretariat thereof accordingly. In any case, the competent authority shall adopt all the necessary measures to ensure the continuity of the activities already undertaken by the technical service.

12. Records on designated technical services

12.1. The competent authority shall maintain records on technical services to demonstrate that requirements for designation, including competence, have been effectively fulfilled.

12.2. The competent authority shall keep the records on technical services secure to ensure confidentiality.

12.3. Records on technical services shall include at least:

(a) Relevant correspondence;

(b) Assessment records and reports;

(c) Copies of designation certificates.